Forgotten Battles of World War II

DR. DONAL MCAULIFFE

Forgotten Battles of World War II
Copyright © 2024 Dr. Donal McAuliffe

All rights reserved. No part of this publication may be reproduced, distributed, or transmitted in any form or by any means, including photocopying, recording or other electronic or mechanical methods, without the prior written permission of the author, except in the case of brief quotations embodied in reviews and certain other non-commercial uses permitted by copyright law.

Without in any way limiting the author's [and publisher's] exclusive rights under copyright, any use of this publication to "train" generative artificial intelligence (AI) technologies to generate text is expressly prohibited. The author reserves all rights to license uses of this work for generative AI training and development of machine learning language models.

This is a work of non-fiction. The publisher and the author have made every effort to ensure that the information in this book was correct at press time. And while this publication is designed to provide accurate information in regard to the subject matter covered, the publisher and the author assume no responsibility for errors, inaccuracies, omissions, or any other inconsistencies herein, and hereby disclaim any liability to any party for any loss, damage, or disruption caused by errors or omissions, whether such errors or omissions result from negligence, accident, or any other cause.

Cover photo courtesy of National Archives Catalog
National Archives Identifier: 531424
Local Identifier: 111-SC-407101
NAIL Control Number: NWDNS-111-SC-407101
For more information: https://catalog.archives.gov/id/531424

Printed in the United States of America

Paperback ISBN: 979-8-9877116-4-4

DartFrog Blue is the traditional publishing imprint of DartFrog Books, LLC.

301 S. McDowell St.
Suite 125-1625
Charlotte, NC 28204

www.DartFrogBooks.com

Contents

Introduction 1
The Battle of the Bzura 5
The Winter War 13
The Battle of Hannut 21
The Battle of Beda Fomm 29
The Battle of Bryansk 37
The Invasion of the Philippines 45
The Battle of the Kerch Peninsula 53
The Aleutian Islands Campaign 61
The Battle of Algiers 69
Operation Gunnerside 77
Black May 85
The Battle of the Dnieper 93
Operation Achse 101
The Battle of Narva 109
The Battle of Kohima 117
The Battle of Hengyang 125
The Operation Dragoon Landings 133
The Liberation of Belgrade 141
The Battle of the Scheldt 149
The Battle of the Leyte Gulf 157
The Battle of Mutanchiang 167

Introduction

It is strange, the manner in which our basic understanding of the Second World War–the defining conflict of the modern world–is shaped by Hollywood. For example, the evacuation of the British Expeditionary Force from Dunkirk in late May and early June 1940 is seen, at least in Hollywood's view, as the defining episode of the Battle of France. The first film treatment of the Dunkirk evacuation, starring John Mills, Bernard Lee, and Richard Attenborough, appeared in 1958, while more recently, this same battle was the subject of a Christopher Nolan film in 2017. The Battle of Britain that followed has also been covered in several motion pictures, from 1969's *The Battle of Britain* to Gary Oldman's Academy Award-winning turn as Winston Churchill in 2017's *Darkest Hour*. The Academy Award-winning biopic of General George S. Patton in 1970 featured Operation Husky, the Allied invasion of Sicily in the summer of 1943, as its central action, while the D-Day landings have been the subject of a great many films, from 1962's *The Longest Day* to 1998's *Saving Private Ryan*. Beyond the efforts of the Western Allies in France and Italy, the Battle of Stalingrad on the Eastern Front and the battles of Midway and Iwo Jima in the Pacific Theatre have also been consistently regarded as important battles, as has, of course, the Japanese attack on Pearl Harbor.

It is a neat running order: Dunkirk, the Battle of Britain, Pearl Harbor, Midway, Stalingrad, Sicily, D-Day, and Iwo Jima, followed finally by Berlin and the atomic bombs. This narrative isn't, strictly speaking, incorrect. These were undeniably some of the most important moments in the Second World War. Yet there were hundreds of major battles and thousands of military engagements in the conflict, the vast majority of which have been ignored by Hollywood and forgotten by most everyone other than the experts working on the history of the war. Clearly not all of these skirmishes can be retrieved from neglect in the pages of one book, but there is certainly room to try to shed light on some of these overlooked battles of the Second World War, which is the purpose of this volume.

Many of the battles showcased in this book were just as important in determining the outcome of the war as those that have been given the Hollywood treatment. Consider, for instance, the fact that the German Wehrmacht did actually make it to the outskirts of Moscow in the early winter of 1941. This was at a time when the Soviet dictator, Joseph Stalin, was giving serious consideration to seeking an agreement with the Nazis whereby he would cede a huge chunk of territory to the Germans in return for peace. Given this, the Battle of Bryansk, discussed in the pages that follow, takes on a particular importance. At Bryansk, some 400 kilometers southwest of Moscow, the Red Army finally managed to put up a spirited defense against Heinz Guderian's Panzer divisions and the German 2nd Army in October 1941. This slowed the German advance on Moscow enough to prevent the Nazis from taking the Russian capital before winter set in. Even though Bryansk resulted in a defeat for the Soviets, had it not occurred, the Russians might well have capitulated, and the entire course of the war and modern history would have been very different.

INTRODUCTION

This is just one example of the importance of the "forgotten" battles and engagements charted in this book. In presenting these twenty-plus battles, a selection has been made from the full six-year span of the war, with care given to cover each of the major fronts in Western Europe, Eastern Europe, North Africa, and the Pacific, as well as the often overlooked fronts in places like Finland, China, Burma, and the Balkans. Indeed, some of these fronts were the most consequential fronts of all, where actions shaped the world order that emerged after the war, and each deserves our attention. The Balkans in Southern Europe, for instance, were dominated in the second half of the twentieth century by Josip Broz Tito, the Yugoslav partisan who emerged in the early 1940s as the anti-Axis leader there and led the liberation of Belgrade in the autumn of 1944. The Balkans, as we will see, are the site of one of the many "forgotten" battles that had resonance well beyond 1945.

The Battle of the Bzura
(9–19 September 1939)

This was the foremost clash of the German invasion and conquest of Poland, involving 425,000 German troops and over 200,000 Poles. The battle's wider significance lay in emphasizing how much warfare had changed since the First World War, and it was particularly notable as the first major demonstration of the German Blitzkrieg, or "lightning war."

The Second World War began in the first hours of September 1939, when, after a false flag operation designed to halfheartedly allow the Germans to claim they had been provoked into attacking Poland, the German Wehrmacht crossed the eastern border into the Polish Republic. Two days later, Britain and France responded by declaring war on Nazi Germany. Though it seems somewhat peculiar to make this distinction, the war was actually preceded by a period of unusually long peace in Europe. The continent had been dogged by war for centuries, there almost never having been an occasion in which a conflict of one kind or another wasn't underway somewhere. Consider, for example, that Spain and the Dutch Republic were at war for eighty years between 1568 and 1648, and the Great Northern War between Russia, Sweden, and various other powers continued for over two decades, from 1700 to 1721. Such interminable conflict was absent from Europe, however, for much of the 1920s and 1930s. With the exception of the Spanish Civil War (1936–1939), there was a period of almost unblemished peace across Europe beginning in 1923, when the revolutions and civil wars that had followed the First World War came to an end. During that long, peaceful period from 1923 to 1939, remarkable technological advances were made. Tanks and airplanes, for example, developed massively from what they had been when the First World War ended in November of 1918. Thus, when Germany invaded Poland in September of 1939, the nature of warfare had changed enormously. Nowhere was that more evident than at the Battle of the Bzura, the largest clash of the entire Polish campaign.

The Battle of the Bzura was fought around the River Bzura, a tributary of the larger River Vistula in what is now central Poland. It was there, just over a hundred kilometers to the west of the capital Warsaw and north of the industrial city of Lodz, that a substantial portion of the Polish army, under the command of General Tadeusz Kutrzeba, launched a counterattack against the Germans on the 9th of September 1939. By that time, the Germans had already progressed well into Poland. It was not the case, however, as is often assumed, that the Nazis faced a small army that was easily defeated. The Poles had nearly a million men in arms, but they faced an enemy that was far more sophisticated, with better generals, heavy guns, brutally fast tank divisions, and one of the most advanced air forces in the world. The Poles had virtually none of these, and the western city of Krakow fell as early as the 6th of September. By that time, German anti-aircraft weapons and fighter planes had completely nullified the small Polish air force and the German *Luftwaffe* (which literally means "air force") had control of the Polish skies. This was the context in which the two sides would clash at the Battle of the Bzura.

General Kutrzeba and his subcommanders led a division of the Polish army known as the Army of Poznan, as well as parts of the Army of Pomorze, at the Battle of the Bzura. This force consisted of just over 200,000 troops. The Poles faced a German army twice that size commanded by Field Marshal Gerd von Runstedt, a veteran German general who had served as a senior staff officer in the First World War and who, at sixty-three years of age, was one of the most experienced generals in the German *Wehrmacht*, the name for the land forces of the German army, or "*Heer*." However, the tactical advantages von Runstedt enjoyed as he advanced towards the River Bzura went beyond numerical superiority. The Germans also had divisions of their Panzer tanks, the central component of what would come to be known as the *Blitzkrieg*, or "lightning war," to compliment the aforementioned air superiority. It was a sign of how ill-prepared the Poles were to meet a modern army of this kind that Kutzreba still had cavalry divisions in the Army of Poznan. These weren't armored divisions euphemistically referred to as "cavalry;" they were literally brigades of men fighting on horseback against a modern army.

The Battle of the Bzura is roughly divisible into two phases. Between the 9th and 15th of September, the Poles launched and sustained their counterattack against the might of the German Wehrmacht, which seems, in retrospect, to have been a bit foolhardy. We might think that the Poles would have been better off using their limited resources to fight a gradual defensive withdrawal to the east, one that would cost the Germans as many men as possible. After all, as military tacticians from Sun Tzu to Carl von Clausewitz have made clear over the centuries, it is always best to be on the defensive. However, it is worth noting that in September of 1939, the Poles were expecting aid from the British and French and were severely misinformed about Franco-British preparedness for war with Germany. Thus, the Battle of the Bzura was entered into by General Kutzreba in the hope that the Poles could push the Germans back into western Poland and keep them there until an expeditionary force arrived from Western Europe.

The counterattack began on the night of the 9th of September as the Army of Poznan attempted to take advantage of a gap in the German lines as the German 8th Army advanced towards the cities of Lowicz and Strykow. The initial engagement was successful, with a few thousand German soldiers either killed or captured and a number of small towns and villages along the southern end of the River Bzura retaken by the Poles. The Germans had been caught unawares and continued to experience difficulties with the Polish advance on the 10th and 11th of September, but they soon managed to regroup. By the 12th of September, divisions of the German 10th and 4th armies were being sent southwards to plug the gap in the German lines near the River Bzura. The Germans also sent Luftwaffe flights over the River Bzura, continuously impeding the advances of Kutzreba's forces. However, the Poles did continue to make piecemeal gains, and by the 14th of September, they had crossed the River Bzura near the city of Lowicz and were preparing to strike at the German positions there. This was the juncture, however, at which the Battle of the Bzura began to turn in the Germans' favor.

On the 15th of September, Kutzreba's forces went on the defensive. The Germans had regrouped around the Bzura, and Kutzreba was receiving intelligence that the German 4th

Panzer Division was rerouting away from its drive towards Warsaw and was now headed for the Bzura, as well. The 1st Panzer Division was already in place near the Polish forces, and on the 16th of September, it and several infantry divisions counterattacked the Polish Army of Poznan. The result was a disastrous flight back over the River Bzura by the Poles, one in which much of their heavy equipment and artillery was abandoned. By the 17th, the Germans were firmly on the offensive and began pushing the Poles back from the east bank of the River Bzura. On the 19th, with their ammunition and food rations running low, Kutzreba and his sub-commanders began a tactical withdrawal towards Warsaw.

The Battle of the Bzura was the largest engagement of the Polish campaign. Despite the Poles being the initial aggressors through their counterattack, the Germans gained the upper hand, and by the end of the fighting twelve days later, some 20,000 Poles were dead, with a larger number wounded and tens of thousands captured as POWs. The Germans lost only 8,000 men, with minimal damage to their tank divisions and artillery units. Overall, the clash underscored the tactical, logistical, and technical superiority of the Wehrmacht over a military that was stuck in a time warp, the Poles having failed to adopt modern methods of warfare in the interwar period.

In the aftermath of the Battle of the Bzura, the Germans continued their swift eastward advance. The Poles were dealt a further crushing blow, even as the battle was still underway in central Poland, when the Soviet Union invaded eastern Poland on the 17th of September. The Germans and Russians had agreed to divide up Poland between them as part of their Treaty of Non-Aggression, an agreement made between the German foreign minister, Joachim von Ribbentrop, and the Soviet foreign minister, Vyacheslav Molotov, on the 23rd of August 1939. General Kutrzeba made his way to Warsaw by the 22nd of September, and there found the Polish military preparing to enter talks whereby they would surrender to the Germans and Soviets; the Polish Prime Minister, Felicjan Sławoj Składkowski, had already fled south to Romania.

Warsaw fell to the Germans on the 28th of September, and on the 6th of October, the invasion of Poland ended with the complete occupation of the country, though no Polish government ever surrendered. The Soviets annexed the eastern parts of the country, while the Germans annexed much of the west and formed the central and southern parts of the country into the General Government of Poland. Overall, the invasion, and the Battle of the Bzura as the main clash within it, clearly demonstrated how much warfare had changed since 1918 and how fast the German Wehrmacht could overrun a large country like Poland. These were lessons the British and the French would soon learn, to their dismay, in northern France. Meanwhile, Poland would become the scene of some of the Nazis' worst crimes over the next five years, with each of the six "death camps" where the Holocaust was carried out–Auschwitz, Treblinka, Sobibor, Belzec, Majdanek, and Chelmno–located there.

SOURCES AND FURTHER READING:

Gorodetsky, Gabriel. "The Impact of the Ribbentrop-Molotov Pact on the Course of Soviet Foreign Policy." *Cahiers Du Monde Russe Et Soviétique 31*, no. 1 (1990): 27-41. https://doi.org/January-March 1990.

Schiman, Frederick L. "The Nazi Road to War:I. The Conquest of Poland." *Current History 24*, no. 137 (1953): 22-25. https://doi.org/January 1953.

Seidner, Stanley S. 1975. *Marshal Edward Śmigly-Rydz and Poland: 1935-1939*. New York: St. John's University.

Zaloga, Stephen J., and Victor Madej. 1991. *The Polish Campaign: 1939*. New York: Hippocrene Books.

Zaloga, Stephen J., and Howard Gerrard. 2022. *Poland 1939: The Birth of Blitzkrieg*. New York: Bloomsbury Publishing.

The Winter War
(30 November 1939 – 13 March 1940)

Surely one of the most forgotten parts of the Second World War was the Soviet invasion of Finland at the end of November 1939, in fulfillment of one of the terms of the Molotov-Ribbentrop Pact between the Soviet Union and Nazi Germany. The war became a disaster for the Russians, as the Finns put up a spirited fight and forced the Soviets into a face-saving peace agreement in March of 1940. This Winter War also gave birth to the term "Molotov Cocktail," the tank-stopping incendiaries the Finns produced and utilized in large numbers and named after the Soviet Foreign Minister, Vyacheslav Molotov.

Something that is often forgotten in the story of the Second World War is that the main conflict in the winter of 1939 and the early spring of 1940 didn't involve the Germans at all. Instead, the main conflict of that period occurred between the Russians and the Finns. As part of the Molotov-Ribbentrop Pact between the Nazis and the Soviets in August 1939, it was determined that, in addition to dividing up Poland between them, the Russians would annex the Baltic States and Finland. This would allow Joseph Stalin to reverse what many of the Soviet leaders viewed as a historical wrong resulting from the Russian Civil War twenty years earlier, when Finland and places like Estonia and Latvia had been able to break free of Russian rule after a century or more of belonging to the Russian Empire. As part of that process, Finland, which since 1809 had been ruled as the Grand Duchy of Finland and part of the Russian Empire, had finally declared itself to be a republic in 1919, and had remained independent of Russian rule for the next two decades. At the end of November 1939, with the Polish campaign over, Stalin ordered the invasion of Finland, with the goal of reconquering a region which he believed rightfully constituted part of the USSR.

The invasion of Finland began on the 30th of November, when nearly half a million Soviet military personnel became active along the Finnish border and the Soviet air force initiated a bombing campaign against the Finnish capital, Helsinki, and other military targets. In response, most of Finnish society mobilized and put up a spirited defense. They needed to. Finland had only 3.7 million people in 1939, many of them women, children, and the elderly, so even fully mobilized, the Finns faced a daunting task. Moreover, the Soviets had thousands of tanks and aircraft, compared to just a few dozen Finnish tanks and an air force of just over a hundred planes, many of them little more than small reconnaissance craft. What the Finns did have, however, was local knowledge of the terrain, and a cause worth fighting for.

The most intense fighting in December 1939 and into January and February of 1940 was concentrated along the border between the Gulf of Finland and Lake Ladoga in the Karelia region. The Finns pulled back to the Mannerheim Line, a defensive line named after Field Marshal Baron Carl Gustaf Emil Mannerheim that the Finnish government established in the 1920s and 1930s to take advantage of the region's many lakes between the Gulf of Finland and Lake Ladoga. Finland is also heavily forested, so with Soviet divisions having to proceed through narrow pathways between the lakes, forests, and fortified sites, the Finns engaged in guerilla warfare, bursting out of the woods to attack Soviet divisions with incendiaries, and even using crowbars and logs to jam the wheels of Soviet tanks and armored vehicles.

This is not to suggest the Soviets were being repelled, however. They consolidated their control over territory south of the Mannerheim Line in December 1939 and the first weeks of 1940, and also sent a larger military contingent than the Finns had been expecting into Lapland in the north, allowing for some advances there. However, the Soviets were held back in other respects, notably losing the protracted Battle of Suomussalmi in the north in January 1940 after a six-week clash. Part of the Russian difficulty was an unusually cold winter, much like the Germans would experience in 1941, the coldest Russian winter of the twentieth century. The winter weather definitely favored the Finns as 1939 became 1940.

The Soviets did begin to make a breakthrough in February 1940, as the limited manpower of the Finns started to suffer from sheer exhaustion. Weeks of being embedded in freezing cold temperatures in guerilla units in the forests and lake regions took its toll, and in the first weeks of spring, the Soviets made advances on a number of fronts, eventually penetrating the Mannerheim Line in early March 1940. That breakthrough would act as a catalyst for both sides to consider peace negotiations, as Stalin was now in a position to demand some territorial concessions and save face at home, while the Finns knew things would get worse if they did not exit the conflict as quickly as possible.

One of the most notable features of the Winter War is its role in the history of the Molotov Cocktail. It was not a new idea in

1939 to craft a makeshift incendiary using commonplace flammable liquids inside a bottle. Millennia earlier, so-called "Greek Fire," a substance which probably involved naphtha and quicklime, had been used by the Greeks and later the Byzantines to destroy enemy ships at sea, and the Chinese had invented a primitive grenade using gunpowder as early as the eleventh century. But the Winter War saw the first systematic use of Molotov Cocktails. These are not complex weapons. A glass bottle is filled with substances like napalm, alcohol, or petroleum. A large strip of cloth or a handkerchief is then stuffed into the top of the bottle. This serves two purposes: it prevents the liquid inside the bottle from spilling out when the incendiary is thrown, and it also acts as a wick. The wick is lit, and the bottle is thrown. The idea is that the glass bottle will shatter on impact, spraying the flammable liquid, which is immediately ignited by the lit wick.

The Finns began using these in large numbers in December 1939 in order to combat Soviet tanks. Because the Russian invasion came about as a result of the Molotov-Ribbentrop Agreement, the Finns developed an ironic name for these rudimentary weapons: the Molotov Cocktail. It was common at the time to declare that you were sending the Russian foreign minister a cocktail to have with his meal. Molotov Cocktails proved extremely effective, particularly once a design was established and the cocktails entered mass production at the Rajamaki distillery north of Helsinki. The Finns ambushed Russian divisions and peppered their tanks with Molotovs from close distance. Over 2,000 Russian tanks were lost during the Winter War, an enormous figure, with many falling prey to Molotov Cocktails, though the figures are disputed as the Soviets tried to understate their losses afterward.

It wasn't just the sheer number of tanks crippled by Finnish Molotov Cocktails that led Stalin and the rest of the Politburo to rethink their strategy in Finland. By the early spring of 1940, upwards of 100,000 Soviet troops were either dead or missing, with nearly twice that number wounded or out of action in some other way. Many men were experiencing frostbite and other conditions associated with the harsh winter of Northern Europe. This was not a condition that would only affect the Germans in 1941 and 1942; a substantial proportion of any Soviet army was made up of troops from places like Kazakhstan, Georgia, and other parts of the Caucasus and Central Asia, men who were not used to such extreme winter conditions. As a consequence of these calamitous losses in men and tanks, the Soviets entered peace talks with the Finnish government in early March 1940.

The Winter War came to an end at midday on the 13th of March 1940, following the signing of the Treaty of Moscow. Under the terms of the Treaty, Finland made land concessions to the Soviets, though it was hardly the total victory Stalin might have wished for. A chunk of land in the Karelia region of southeastern Finland was ceded to Russia, giving the USSR complete control of Lake Ladoga, where Finland had occupied the northern shore since 1919. Some islands in the Gulf of Finland also came into Russian possession, as did a relatively valueless strip of land at Salla in northern Finland along the Russian border. This constituted just under one-tenth of Finnish territory, as well as about one-eighth of the Finnish population; a notable loss, though hardly a cataclysmic defeat for a small nation given the odds it faced when the Winter War began at the end of November 1939.

The Winter War was important in pointing towards the fundamental weakness of the Red Army at the start of the Second World War, with its numerical superiority and abundance of resources counteracted by a more effective and motivated small Finnish army. This was not a new phenomenon. The Russian military had been underperforming since the Russo-Japanese War of 1904 to 1905, when a European power was defeated by a non-western state in a conflict for the first time in the modern era. In 1917, the abysmal performance of the Russian Imperial Army on the Eastern Front against the Germans led to the collapse of the Tsarist government, while the lack of sophistication of the Red Army in combating the Japanese in the undeclared border war around the Mongol and Manchurian border between 1935 and 1939 again pointed to the inadequacies of Russian leadership and tactics. Yet for all the lessons that should have been learned from the Winter War, the Soviets failed to reform their military in 1940, and when the Germans invaded the USSR in June 1941, the Red Army would experience a series of catastrophic defeats in summer and autumn.

The British and French had attempted to send aid to the Finns during the Winter War. The plan had been to dispatch an expeditionary force to the port of Narvik in Norway, which would then proceed overland across Norway and Sweden and into northern Finland. The Allied plan was to provide ancillary support to the Finns while avoiding direct clashes with the Soviets, an approach which in some ways mirrors NATO's support of Ukraine against Russia in the current war. However, gaining permission to send troops through Swedish territory was a vexatious issue, primarily because while Sweden successfully maintained its neutrality during the Second World War, it was broadly more favorable to the Nazis for the first few years of the conflict. Thus, the Allied aid to Finland never materialized. However, Soviet intelligence miscalculated the degree to which British and French aid to the Finns was imminent, and that played a role in Moscow's decision to sue for peace in March 1940.

The Winter War had a broader significance for Finland's role in the Second World War going forward. Again, it is often completely overlooked in general histories of the war that Finland joined the Nazis in their invasion of the Soviet Union in the summer of 1941. This was an act of revenge for the Winter War on the part of the Finns, as they saw an opportunity to reclaim the land they had lost in March 1940. The Continuation War, as it is known, began on the 25th of June 1941, three days after the commencement of the German invasion of the USSR through Operation Barbarossa. The Finns hoped to create a Greater Finland by reconquering territory that had been lost to the Russians centuries earlier and extending Finnish territory eastwards to the White Sea. Over the next three years, the Finns contributed hundreds of thousands of troops to the war effort, notably to the Siege of Leningrad. This made the Finns the second most important ally of Germany in the invasion of the USSR, second only to Romania. As the war effort turned in 1943 and 1944, the Finnish government saw the writing on the wall fairly quickly, however, and agreed to the Moscow Armistice with the Russians in September 1944. Through this agreement, the two sides largely returned to the borders agreed on through the Treaty of Moscow in March 1940 at the end of the Winter War. One of the major features of these events was that Finland remained independent of the Soviet bloc after the Second World War.

SOURCES AND FURTHER READING:

Anderson, Albin T. "Origins of the Winter War: A Study of Russo-Finnish Diplomacy." *World Politics* 6, no. 2 (1954): 169-189. https://doi.org/January.

Gorodetsky, Gabriel. "The Impact of the Ribbentrop-Molotov Pact on the Course of Soviet Foreign Policy." *Cahiers Du Monde Russe Et Soviétique 31*, no. 1 (1990): 27-41. https://doi.org/January-March 1990.

Kolomyjec, Maksim. 2011. *Tanks in the Winter War, 1939-1940*. London: Amber Books Ltd.

Sander, Gordon F. 2013. *The Hundred Days Winter War: Finland's Gallant Stand Against the Soviet Army*. Lawrence, Kansas: University Press of Kansas.

Trotter, William R. 2002. *The Winter War: The Russo-Finish War of 1939-1940*. London: Aurum Press.

The Battle of Hannut
(12–14 May 1939)

The story of the German invasion of the Low Countries and France in May 1940 is usually simplified to suggest that the Germans invaded the region, pushed the British Expeditionary Force all the way to the English Channel, and then captured Paris after the British evacuation at Dunkirk. However, there was concerted fighting during the Battle for France, notably at the Battle of Hannut, the largest tank battle in history when it occurred in mid-May 1940 in Belgium.

While the Soviets were unsuccessfully trying to conquer Finland, the Nazis were surprisingly quiescent after conquering Poland in the autumn of 1939. The French had made an often overlooked effort to invade the Saar region of western Germany in September 1939, but they were quickly repelled, and thereafter the British and the French focused their efforts on rearming as quickly as possible, surprised by the speed and ease with which the Germans had conquered Poland. The Germans, too, took their time expanding the conflict after their eastward expansion to Warsaw, and with months of inaction on all sides, there was growing talk by the spring of 1940 about a "Phoney War." Anyone who was wistfully hoping for an abortive conflict was soon disabused of such a notion, however, as the Nazis quickly took control of Denmark and Norway in precision campaigns beginning in April 1940. Both Denmark and Norway were neutral, but their strategic positions on the North Sea led the Germans to occupy them; the British had been planning a similar tactical occupation of Norway.

Following the swift campaigns into Scandinavia, the focus of Nazi military activity shifted towards France and the Low Countries. The Battle of Hannut would constitute one of the first major engagements there. The German plan for invading France was arrived at after much deliberation with the high command of the German Wehrmacht, in consultation with Hitler and other members of the Nazi regime, such as the head of the Luftwaffe, Hermann Goering. The generals were divided about how to proceed, with some favoring an approach that would involve a full-scale invasion of Belgium, steering a more northerly course, while others, notably General Erich von Manstein, arguing that the main prong of the invasion should move through the Ardennes region of southern Belgium and Luxembourg and then proceed quickly into northeastern France. This military preparation was further complicated by the Mechelen Incident in January 1940, when a German plane carrying papers that revealed parts of the Nazi war plan crashed near Mechelen in Belgium. With all of these factors taken into account, the plan that was eventually decided upon was contained in *Fall Gelb*, or "Case Yellow." It was decided that the Wehrmacht would proceed in two main divisions. One, Army Group A, would move through the Ardennes into northeastern France, while the other, Army Group B, would follow a more northerly course through the Low Countries. The latter effort was more of a feint designed to draw Allied troops northwards while the more southerly army drove towards the English Channel, in the hope of trapping large sections of the British Expeditionary Force and the French between the two German forces.

For their part, the French had been preparing since the 1920s for a possible German war of revenge and had erected the Maginot Line of defensive fortifications along their eastern borders. The notion that the French had been foolish enough to have neglected to extend these fortifications along the Franco-Belgian border, despite the Germans having invaded France through Belgium back in 1914, is a myth. The Maginot Line did partially extend along the Belgian border, though it was weakened there owing to Belgian objections in the 1920s and 1930s. Furthermore, some parts of it were weaker in the region next to the Ardennes, for the simple reason that the Ardennes region was so hilly and forested that the French did not expect a German attack there, or even think it was possible in large numbers.

The Battle of Hannut would take place next to the city of Hannut in central Belgium and was part of the feint northwards by Army Group B that was designed to pull French and British forces away from the main German invasion force, which would then strike through the Ardennes region. The wider Battle of France commenced on the 10th of May 1940, the same day that Winston Churchill succeeded Neville Chamberlain as Prime Minister of Britain. Hundreds of thousands of German troops took part in the initial operations and streamed into the Low Countries; the invasion would eventually involve millions of German military personnel. As tank commanders like Heinz Guderian led the swift onslaught through the Ardennes, the Panzer divisions to the north were commanded by individuals like Erich Hoepner and Johann Joachim Stever, who together would command the German lines at Hannut. Guderian and the commanders would not begin their attack until the 15th of May, the plan being that by then, the French and British would have diverted a huge proportion of their forces northwards into central Belgium to block the advance of Army Group B.

The Battle of Hannut began on the 12th of May 1940, the third day of the Battle of France, though the first week of combat took place more in the Low Countries than in France. The battle at Hannut was a relatively evenly matched affair in terms of personnel. Hoepner and Stever had roughly 25,000 men under their command, while the French had marginally fewer men under the command of René Prioux, a veteran French cavalry commander who had served in the First World War. Both sides had approximately 600 tanks and armored vehicles under their command, making this an enormous tank engagement early in the war.

The battle that followed raged for nearly three days. It was an even-sided affair from the very beginning, with the French having significant air support from the British Royal Air Force (RAF), and also being evenly matched in terms of firepower. Prioux was a crafty commander, allowing the Germans Hoepner and Stever to take the town of Hannut early on the 12th, and then tactically outflanking them. As a result, by the end of the first day, the two German commanders were unsure of the exact positioning of the French around the town and were wary of any rapid movement forward. The French tanks and armored vehicles had also held up very well against the technologically more advanced German Panzer tanks.

Fighting the following day focused on the district to the west of Hannut as the Germans tried to break through the French lines. A tank battle to secure the village of Orp ensued, with heavy fire exchanged. Still, no conclusive victory was achieved by either side, and fighting dragged on into the 14th. This was the last day of the Battle of Hannut and resulted in the French retreating to the town of Gembloux. However, this did not mean that the Germans had won. The French had always intended to fall back to Gembloux, and the fighting at Hannut had simply been designed to slow the advance of Army Group B through central Belgium while the French army and British Expeditionary Force regrouped in western Belgium and northeastern France. Moreover, the French managed to destroy over fifty German tanks and damage a hundred more, while sustaining marginally fewer losses themselves. From this perspective, Prioux and the French gained a victory of sorts, a rarity in a conflict which otherwise was defined by superior German tank warfare.

At the time it was fought, the Battle of Hannut, which has sometimes been called the Battle of the Belgian Plain, was the largest tank battle in history, with over 600 German Panzer tanks and a similar number of less powerful and slower French armored fighting vehicles deployed. It isn't hard to see why this eclipsed all previous tank battles. The first tanks had been deployed in September 1916 by the French and British at the Battle of the Somme during the First World War as the various combatants sought to break the impasse of trench warfare on the Western Front. Tanks were used in increasingly large numbers by the French, and to a lesser extent by the British, in 1917 and 1918, while the Germans were surprisingly reluctant to widely embrace them. It was only during the 1920s, and particularly the 1930s, that various European states began establishing entire tank divisions, which were often called "mechanized cavalry." Thus, the Battle of France was the first engagement since the First World War in which large tank divisions clashed, and Hannut was the apex of this. As with so much in the Second World War, the scale of what occurred at places like Hannut in Western Europe would soon be dwarfed by occurrences on the

Eastern Front between 1941 and 1944. The Battle of Bialystok-Minsk in Belarus in the first weeks of Operation Barbarossa, for example, would see over 6,000 tanks deployed by the Germans and Soviets, while in July and August 1943, the Battle of Kursk in western Russia involved 10,600 German and Russian tanks in what is still, by some stretch, the largest tank battle in history.

We might ask what the tactical significance of the Battle of Hannut was in the overall scheme of the Battle of France. It could be said that it made no difference. The day after the tactical withdrawal from Hannut, Army Group A began its movement through the Ardennes Forest and into northeastern France, with Heinz Guderian's tank divisions reaching the English Channel on the 20th of May, a movement that was so fast that the German high command thought Guderian must have been mistaken when he reported this back to Berlin. The Ardennes movement had been so successful that it left the British Expeditionary Force and sizeable sections of the French military cut off from Paris and stranded between Army Group A in the south and Army Group B to the north.

What followed is well-known. Hundreds of thousands of Allied troops streamed into the port town of Dunkirk and only survived because Goering convinced Hitler to stop the Wehrmacht outside the town so that he could order the Luftwaffe to bomb the British Expeditionary Force into oblivion inside Dunkirk and claim credit for himself. Instead, the Miracle of Dunkirk occurred in the final days of May and early June, as over 300,000 troops were evacuated in a giant armada of small British boats and fishing vessels. It could be argued that, in slowing the advance of German Army Group B through Belgium at the Battle of Hannut, the French aided the successful British Expeditionary Force withdrawal at Dunkirk. Meanwhile, the French government fled Paris on the 10th of June. The capital fell four days later, leading to the Nazi occupation of northern and eastern France, Paris, and the Atlantic coastal region. The Vichy collaborationist government was established to control central, western, and southern France.

SOURCES AND FURTHER READING:

Clark, Allen F. Jr. "Operations in Belgium and France, 1940: First Phase-The Battle of Flanders." *The Military Engineer 33*, no. 192 (1941): 448-453.

Frieser, Karl-Heinz. 2013. *The Blitzkrieg Legend: The 1940 Campaign into the West.* Annapolis: Naval Institute Press.

Gunsburg, Jeffrey A. "The Battle of the Belgian Plain, 12-14 May 1940: The First Great Tank Battle." *The Journal of Military History 56*, no. 2 (1992): 207-244. https://doi.org/April.

Pitt, Barrie. 1996. *The Military History of World War II.* Vacaville, CA: Bounty Books.

Powaski, Ronald E. 2002. *Lightning War: Blitzkrieg in the West.* Hoboken, NJ: Wiley.

The Battle of Beda Fomm
(5–7 February 1941)

The North Africa Campaign is generally famed for the two battles of El-Alamein in Egypt, but perhaps the most important clash was early on between the British and Italians at Beda Fomm in Libya. It was a significant victory for the British, in which 25,000 Italian POWs were captured, and resulted in Hitler having to dispatch the Afrika Korps to the Mediterranean under Erwin Rommel and pull otherwise vital German resources away from Europe as the invasion of Russia was being prepared that spring and summer.

In the aftermath of the Battle of Hannut, the evacuation from Dunkirk, and the conquest of France in the summer of 1940, the Battle of Britain began. For the better part of a year, the Germans focused on trying to bludgeon Britain into submission by cutting off its resources through submarine warfare in the North Atlantic and psychologically terrorizing the British people through a bombing campaign known as the *Blitz* (which literally means "lightning"). Much to Hitler's dismay, the British did not simply capitulate and join the German military alliance in advance of his invasion of the USSR, something which the Nazi leader had viewed as likely and in line with his racial view of the world. Galvanized by Winston Churchill's wartime leadership, and with aid arriving from a still-neutral United States through the Lend-Lease program, the British fought on.

Without any hope of a British counterattack into Western Europe in 1940 or 1941, much of the focus of the war shifted to North Africa. The fighting there was initially between the British and the Italians, and this led in due course to the Battle of Beda Fomm. Though it is an underappreciated fact, fascist Italy had displayed considerable reservations in allying with Nazi Germany. As late as the Munich Conference of September 1938, where the fate of the Sudetenland in Czechoslovakia was decided, the Italian fascist leader, Benito Mussolini, was more inclined to side with the British Prime Minister, Neville Chamberlain, and the French Prime Minister, Édouard Daladier, in opposing Hitler's expansionism. Additionally, while Italy subsequently did ally with Germany, the Italians did not, as is sometimes assumed, automatically join the war in September 1939. It was not until the 10th of June 1940, when France was effectively defeated, that Mussolini finally entered the war on Germany's side. He did so primarily out of a desire to create a new Italian Mediterranean empire, one which would rebuild the old Roman Empire. A lot of his ambitions focused on North Africa. Mussolini wanted to expand Italy's longstanding colony of Libya to conquer British Egypt, and in the process, unite Italy's Libyan colony to its lands in the Horn of Africa, where the Italians had conquered Abyssinia in the Second Italo-Ethiopian War of 1935-1937. Hitler encouraged his ally's ambition, in part because he wished to secure the Suez Canal and potentially expand Axis influence into the Middle East.

 The Battle of Beda Fomm was one of the major engagements of Operation Compass, a British campaign initiated in December 1940 to withstand Italian efforts to encroach into Egypt from Libya, and if possible, to push the Italians westwards towards Tunisia. At this early juncture in the war, there was no real strategic design to conquer North Africa from the

Axis, Libya being in Italian hands and Tunisia, Algeria, and Morocco being controlled by the collaborationist government of Vichy France. However, Churchill wanted to ensure that Suez was well protected, and he also wished to score a military success that might bolster public morale back home in Britain, where the Blitz was underway. Beginning in mid-December 1940, the Western Desert Force under General Archibald Wavell and Lieutenant-General Richard O'Connor, as well as detachments of the Australian 6th Division, a Commonwealth battalion that arrived in Egypt through the Red Sea, pushed into western Egypt. Though initially consisting of not much more than 30,000 men, they made swift advances against the numerically superior Italian 10th Army, pushing them back into eastern Libya by January 1941. Much like the Russians in Finland, the Italians performed dreadfully against the British, and by the start of February were retreating backward into central Libya near the city of Benghazi.

The Battle of Beda Fomm took place on the 6th and 7th of February 1941 at the town of that name (Bi'r Bayda' Fumm in Arabic), which lies in the Cyrenaica region of Libya about 80 kilometers to the south of Benghazi. It involved elements of the Italian 10th Army under Rodolfo Graziani and a British-Australian contingent under Wavell and O'Connor. Wavell and O'Connor were also supported by Combeforce, or Combe Force, a flying column that derived its name from the fact that it was commanded by Lieutenant-Colonel John Combe. Combeforce would be crucial in what followed, as it was initially given the responsibility of trying to intercept the 10th Army as it attempted to retreat south and then west along the coastal road from Benghazi, which was known by its old Roman name, *Via Balbia*.

The Battle of Beda Fomm was preceded by a movement by Combeforce to stop the Italian 10th Army from proceeding south. Wavell and O'Connor's men were slowed by the rugged terrain east of Benghazi, and by the Western Desert more generally, and could not proceed quickly enough to stop the retreating Italians. Consequently, Combeforce was sent ahead to set an ambush of mines along the *Via Balbia* at Sidi Saleh to the south of Benghazi, on the way to Beda Fomm. On the 5th of February, they set the mines, and just half an hour later, the Italians arrived. The first few exploded vehicles destroyed the road and blocked Graziani's forces from advancing further south, keeping the Italians stuck overnight on the 5th, while Wavell and O'Connor advanced towards the Beda Fomm region from the east.

The Battle of Beda Fomm continued into the morning of the 6th of September as the Italians, numbering at least 25,000 men, tried to fight their way out of the ambushed region against a substantially smaller force commanded by Combe. Combeforce's goal, though, wasn't to try and defeat the Italians there, but rather to prevent them from breaking out of their position long enough for Wavell, O'Connor, and the joint British-Australian forces to reach the Beda Fomm region. They made good strides towards achieving this, with the British reaching Benghazi on the 6th and occupying the city with almost zero resistance. Indeed, the British were broadly welcomed by the local Libyans, who loathed the Italians after decades of extensive colonization of Libya, during which the Italian government engaged in quasi-genocidal policies towards the Libyan people.

With Benghazi under their control, O'Connor immediately sent additional divisions south to Beda Fomm on the 6th. There

THE BATTLE OF BEDA FOMM

was considerable distance to cover and the British would not arrive until the 7th, but the Italians were aware that they would be surrounded, with the sea to the west, the desert to the east, the British and Australians from Benghazi advancing to the north, and Combeforce to the south. Consequently, the Italians attempted to use their few dozen tanks to force a breakout just before dawn on the 7th. They managed to make a bit of progress, but even as they were fighting to try and force a breakthrough, with more and more Italian tanks being knocked out by anti-tank guns, the Italians could hear the British tanks advancing from the north. Once O'Connor's troops arrived from the north in large numbers, the Italian commander on the ground, Giuseppe Tellera, who was badly wounded and would die later that day, ordered the surrender. By the time the Italians surrendered, over a hundred of their tanks and vehicles were burnt along a fifteen kilometer stretch of road, from where the initial ambush at Sidi Saleh had been set and south towards Beda Fomm. Approximately 25,000 Italians of the 10th Army were taken as prisoners of war.

The Battle of Beda Fomm was an enormous morale boost for the British and their few allies at a time when the Germans were generally having their way in Europe. The Battle of Beda Fomm also had wider implications for the whole conflict, because after the calamity that Operation Compass proved to be for Italian ambitions in Africa, Mussolini was forced to appeal to Hitler for military aid. This led to the establishment of the famous *Afrika Korps* under the German commander Erwin Rommel. This expeditionary force of German soldiers had been created in January 1941 and was in the process of being assembled, but Rommel was only appointed its commander four days after the defeat at Beda Fomm. Arriving in North Africa weeks later with over 30,000 German troops, tanks and armored vehicles, Rommel soon demonstrated how superior tactics and training were central to the war effort. Where Graziani's 10th Army of approximately 150,000 men had completely buckled against a numerically inferior British and Australian force in late 1940 and early 1941, Rommel's Afrika Korps went on the offensive. Benghazi was retaken by the Axis powers at the beginning of April, and the Germans and Italians were soon pressing back into western Egypt, with Suez once again under threat. It would take the US entering the war before the Allies could turn the tide decisively in North Africa, but the distraction of North Africa, and the diversion of resources there in the spring of 1941, meant that these units were not available for the invasion of Russia. As we will see in discussing the Battle of Bryansk, the success of the German invasion of the USSR in the autumn and winter of 1941 was a matter of small margins.

SOURCES AND FURTHER READING:

Arthur, Douglas. 2000. *Desert Watch: A Story of the Battle of Beda Fomm.* North Yorkshire, UK: Blaisdon Publishing.

Carrier, Richard. "Some Reflections on the Fighting Power of the Italian Army in North Africa, 1940-1943." *War in History 22*, no. 4 (2015): 503-528.

Macksey, Kenneth. 1972. *Beda Fomm: The Classic Victory (History of the 2nd World War).* New York: Ballantine Books.

Sadkovich, James J. "Understanding Defeat: Reappraising Italy's Role in World War II." *Journal of Contemporary History 24*, no. 1 (1989): 27-61.

Sadkovich, James J. "Of Myths and Men: Rommel and the Italians in North Africa." *The International History Review 13*, no. 2 (1991): 284-313.

The Battle of Bryansk
(2–21 October 1941)

> The German invasion of the Soviet Union, and the speed with which the Wehrmacht advanced into Belarus, Ukraine, and then Russia in the autumn of 1941, was striking. Accordingly, the Battle of Bryansk in October was pivotal in galvanizing Soviet resistance and slowing the Nazi advance on Moscow. Although the Nazi Panzer divisions under tank commander Heinz Guderian ultimately defeated the Red Army at Bryansk, by slowing the German advance there for several weeks before the Russian winter set in, the battle prevented Moscow from falling in the winter of 1941.

While the Afrika Korps were heading across the Mediterranean to Libya, a strategic *volte-face* was being planned in Berlin. The Blitz and the naval blockade of Britain had failed to quickly bring the British to the negotiating table as the Germans had hoped. Instead, Winston Churchill's government had made it abundantly clear that it intended to fight the Nazis to the bitter end. Hitler soon tired of the effort, and, faced with a choice between expending vast resources on attempting a full-blown invasion of Britain or utilizing those same resources against the ideological enemy of fascism–the communists of the Soviet Union–the Nazi leader decided that the latter option was far more attractive. Hence, in the spring of 1941, the Germans began pulling divisions of the Wehrmacht away from Western Europe and sending them east to Poland to prepare for war against the Soviets there. The Non-Aggression Pact signed between Germany and Russia in August 1939 would be dispensed with, and a massive invasion of the Soviet Union would begin. As a result, the bombing campaign over Britain was scaled back enormously, though the naval war, the Battle of the North Atlantic, would continue between the British and the Germans for years.

Operation Barbarossa, the codename for the German invasion of the USSR, began on the 22nd of June 1941. With three million German soldiers and military personnel advancing into eastern Poland, the Baltic States, and western Ukraine, this was the largest military campaign in human history. The goal was to mimic the tactics used in both Poland and France by advancing rapidly and defeating the Soviets before they had time to organize their defense. In order to do so, the main targets were Moscow and Leningrad, which Hitler wanted to secure before the end of 1941. On the way, major cities like Minsk would be captured, while further south, Kyiv was a priority. Effectively, this would be the same kind of Blitzkrieg that had proved so effective in 1939 and 1940, only on a much larger scale.

The Nazi invasion of the USSR met with the kind of success the German Wehrmacht had become used to in the first two years of the war. When Germany advanced into the Soviet Union in the late summer and early autumn of 1941, Vilnius in Lithuania was secured on the third day of Operation Barbarossa, and Minsk fell within a week. In July, hundreds of kilometers of Soviet territory were occupied by the Germans, and enormous numbers of tanks and planes were captured as divisions of the Red Army were surrounded and taken prisoner in huge numbers. This established a pattern of German tactics, whereby they sought to encircle elements of the Red Army and take tens, or even hundreds, of thousands of Soviet troops prisoner at one time.

The German gains continued into August and September. By then the Germans were advancing on three distinct fronts, with one prong of the invasion force heading north towards Leningrad (modern-day St Petersburg), another heading northeast towards Moscow, the ultimate target, and another steering a line straight east to Kyiv. As the Red Army melted away in front of the advancing Germans, it appears that the Soviet government of Joseph Stalin began to ponder negotiating peace terms and surrendering extensive territory to the Germans, with Moscow asking elements within the Bulgarian government to establish potential peace talks. Whether this was a ruse of some sort, or if Stalin actually considered it, remains a point of debate amongst historians of the Second World War. In any event, the Germans continued to experience enormous victories. At the siege of Kyiv in September 1941, the Germans completely encircled the Soviets, and in taking the city late that month, captured almost half a million Red Army soldiers, who became prisoners of war.

At this juncture, it looked as though cataclysmic defeat was inevitable for the Soviets. But then the Battle of Bryansk took place. At first glance, it might be hard to view the engagement on the Bryansk Oblast as anything other than another total German victory. Yet the Red Army did not need to defeat the Germans there; they just needed to slow them down. The Bryansk Oblast lies between 400 and 500 kilometers southwest of Moscow, and today is the administrative region of Russia that meets the borders of both Belarus and Ukraine. It lay on the southern edge of the route that Army Group Center, the section of the German army tasked with capturing Moscow, was traveling in the final weeks of August and early September 1941, after the capture of Minsk and the Battle of Smolensk further to the north in Belarus. It was important to secure Bryansk in order to protect sections of Army Group Center, which was passing further to the north toward Moscow.

Bryansk became particularly important in the second half of September, as it was through here that divisions of Army Group Center, which had been sent south temporarily by Hitler and the German generals to take part in the successful capture of Kyiv, would have to pass in order to rejoin the main force on the way to Moscow. Thus, commanders of the Red Army, such as General Semyon Timoshenko of the 50th Army and General Andrey Yeryomenko of the 13th Army, began moving hundreds of thousands of Soviet troops into the Bryansk Oblast in late September. The German force in the Battle of Bryansk was much smaller, but the core was the 2nd Panzer Army commanded by Heinz Guderian. Guderian had led the Blitzkrieg to the English Channel in northern France in May 1940 and played critical roles in the capture of Minsk and Smolensk in the summer of 1941, and Guderian and his unit were well-established as the most successful tank division in the German army.

Contrary to the popular notion that the Germans were unprepared for the Russian winter, it was actually central to their planning, and the Germans knew they needed to get to Moscow and capture the city before the worst of the winter weather set in during November and December. Therefore, Guderian and other commanders were ordered to proceed as quickly as possible to break through the Russian wall of troops on the Bryansk Oblast. The offensive began in the final days of September 1941, and coalesced on the oblast on the 2nd of

October 1941. From any statistical and logistical perspective, what followed was bruising for the Red Army, as they were gradually pushed back over the next two weeks and suffered severe losses. Timoshenko might well have decided to continue to hold the line and throw troops at the onslaught of Guderian's tank divisions, but by the 5th and 6th of October, the worry was that the Red Army would be encircled, leading to the capture of hundreds of thousands more Soviet troops. Consequently, a staggered defensive withdrawal continued for two weeks. In the end, hundreds of thousands of Red Army soldiers were killed, wounded, or captured, though it took until the 21st of October 1941 before the Bryansk Oblast was fully secured by the Germans–a critical point.

On the 9th of October 1941, a German Reich press officer by the name of Otto Dietrich gave a press conference in Berlin where, on behalf of the government, he declared that the eastern campaign against the Soviet Union would end within weeks in total German victory, a result of the defeat of *Heeresgruppe Timoshenko*, the name given by the German command to the Red Army forces on the Bryansk Oblast. It must have seemed as though the Battle of Bryansk, which was nowhere near over when Dietrich gave his press conference, was a great victory, but ultimately, it was a costly one. The battle had slowed the Wehrmacht's advance towards Moscow by several weeks at a time when the drive towards the Russian capital had already been delayed by the diversion of troops south to Kyiv. Hence, it was late October before German divisions began traversing the last hundred miles between them and Moscow. The German army was also starting to experience difficulties it hadn't previously experienced in Poland, the Low Countries, France, or in the first months of Operation Barbarossa in Poland, Belarus, and western Ukraine; specifically, the difficulties of overstretched supply lines and sheer attrition through months of combat.

The significance of the Battle of Bryansk and the delay it caused in the advance on Moscow would become apparent in November and December 1941. In November, the Germans advanced close to the Russian capital, with advance units of the Wehrmacht reaching the outskirts of the city, but they stopped as they met intense resistance from the Red Army. Winter was setting in, and the winter of 1941 proved to be the worst of the twentieth century in Russia. The oil in the German tanks froze, the winter clothing issued to the rank and file proved inadequate, and German casualties escalated considerably. If Moscow was the center of the Soviet line, then the center held that winter. Had the Germans reached Moscow a few weeks earlier, things might have worked out very differently.

Hitler and the generals realized that the delay in reaching Moscow in the autumn of 1941 had been a catastrophic error; that was evident from their dismissal of Guderian from his command in the final days of December 1941. The rationale would have been relatively clear to Guderian and others, because despite the victory at Bryansk and around Tula in the weeks that followed, Guderian's advance had been slower than usual, and this had potentially affected the outcome of the Battle of Moscow in November and December. Guderian would later be recalled, but there was a tacit realization in this initial dismissal that the three weeks it had taken to get through the Bryansk Oblast were vital.

THE BATTLE OF BRYANSK

The failure to take Moscow that winter was critical in shaping the outcome of the Second World War. Had the Russian capital fallen, there is a possibility that the Soviet will to resist might have buckled, and the war on the Eastern Front would have ended in German victory in late 1941 and early 1942. Instead, Moscow did not fall, the campaign to seize Leningrad by Army Group North also failed, and, as we will see in a subsequent chapter, the Germans were also running into difficulties in the south in their attempt to secure control of the Crimean Peninsula. By early 1942, the war was becoming a bitter conflict of attrition on the Eastern Front, and with British and American war materiel arriving in Russia in increasing amounts, there was no doubt that the Red Army would eventually emerge victorious. It simply had far greater manpower to draw on.

A final notable point about the Battle of Bryansk is that it was central to the development of the most iconic and widely used assault rifle of the twentieth century. One of the Russian combatants on the Bryansk Oblast in October 1941 was Mikhail Kalashnikov. A man of small height and stature with some engineering skills, he had been assigned as a mechanic to a tank unit in the Red Army to fix the Soviet standard-issue T-34 tanks on the Eastern Front. Kalashnikov was wounded in the fighting at Bryansk, and it was while he was recuperating in the hospital afterward that he developed the idea for the assault rifle that bears his name: the AK-47, meaning Kalashnikov Automatic, plus the number 47 because the most widely made model of the gun was first produced in 1947. Kalashnikov developed the first model after soldiers he met in the hospital told him about the flaws in their standard-issue assault rifles. He decided to design his own weapon, one which was effectively a submachine gun with a gas-fired carbine cartridge. Thus, the Battle of Bryansk was not just important in slowing the German advance on Moscow by a critical number of weeks in October 1941, but it also led indirectly to the development of the AK-47.

SOURCES AND FURTHER READING:

Assmann, Kurt. "The Battle for Moscow, Turning Point of the War." *Foreign Affairs 28*, no. 2 (1950): 309-326.

Dimbleby, Jonathan. 2021. *Barbarossa: How Hitler Lost the War*. New York: Viking Press.

Macksey, Kenneth. 2017. *Guderian: Panzer General*. Chicago: Frontline Book Publishing.

Stolfi, Russell H. "Chance in History: The Russian Winter of 1941-1942." *History 65*, no. 214 (1980): 214-228.

Stolfi, Russel H. "Barbarossa Revisited: A Critical Reappraisal of the Opening Stages of the Russo-German Campaign." *The Journal of Modern History 54*, no. 1 (1982): 27-46.

The Invasion of the Philippines
(8 December 1941 – 8 May 1942)

One could be forgiven for thinking that the Japanese attacked only Pearl Harbor in early December 1941, but this is not so. Within hours of the assault by the Japanese on the Pacific Fleet in Hawaii, a separate Japanese attack on the American protectorate in the Philippines began just after midnight on the 8th of December 1941. The Philippines remained under attack for months, until it finally fell to the Japanese in May 1942.

On the other side of the world from Moscow, the Japanese government was about to make one of the most ill-judged decisions of the twentieth century. Its war with the United States was the product of decades of Japanese expansion in the Pacific. After the Meiji Restoration of 1868, which brought the old feudal government of Japan to an end, the country had modernized in an incredibly effective way and had begun to expand in the Western Pacific, establishing Korea as a protectorate, and then altogether annexing the Korean peninsula into the burgeoning Empire of Japan in 1910. In 1931, Tokyo took advantage of the Chinese Civil War to occupy the Manchuria region of northeastern China, which it formed into the puppet state of Manchukuo. The ambitions of the Japanese were in no way satiated, however. They invaded and conquered the great cities of eastern China–Shanghai, Beijing, and Nanjing–in 1937, and had plans to take both Hong Kong and Singapore from the British and Indochina from the French.

All of this would require resources, particularly oil, and steel, and Japan was running low on both by 1940-1941 as the United States, wary of Japanese expansion around the western Pacific, curtailed its sale of petroleum to Japan. This left the emperor and his ministers in Tokyo in a difficult position. The best available large supplies of oil in the Far East were in the oil fields the British had developed in Burma and Borneo, or in the Dutch colonies of the East Indies. To acquire these, however, would mean going to war with Britain and the United States, the latter of which controlled the Philippines, a barrier to Japanese expansion beyond the South China Sea. The question, then, was simple: would Japan cease its aggressive expansion because of a lack of oil and other vital resources, or would it risk war with the US in order to expand into the East Indies and Burma? As we all know, it chose the latter option, and on the 7th of December 1941, the Japanese launched a surprise assault on the US Pacific Fleet lying at anchor in Pearl Harbor on the Hawaiian island of Oahu.

The Pearl Harbor element of these events is well-known, with the attack leading to a temporary crippling of American sea power in the Pacific, but also dragging America into a war it had been reluctant to enter up to that point. After Pearl Harbor, Washington was committed to a war on two fronts, not only with the Japanese, but also, owing to Japan's alliance with the Germans and Italians in the Axis Powers, in Europe. The more forgotten element of these events is that Pearl Harbor was not the only place the United States was attacked at the very beginning of the war. The Japanese also struck at the Philippines and Guam on the 8th of December, the goal being to occupy the US-held territories on the western side of the

THE INVASION OF THE PHILIPPINES

Pacific and eliminate the possibility of their being used by the US as advance bases against Japan in the war that was beginning. The islands of the Philippine Archipelago were doubly important to Japan, as they could also be used as staging posts for Japanese movement further southwest into the East Indies, and the exploitation of the oil fields there.

Guam fell quickly, but the attack on the Philippines would turn into one of the most bitter engagements of the entire war in the Pacific. Within hours of the strike against Pearl Harbor, forces led by Japanese Lieutenant General Masaharu Homma began attacking the Philippines from Formosa, as Taiwan was known at that time. The initial foray largely consisted of a bombing mission against the American airfield on the northernmost and largest island of the Philippine archipelago, Luzon, where the US had concentrated many of their fighter and bomber planes at Clark Field in the north. The attack resulted in the loss of dozens of B-17 bomber planes and P-40 fighter planes, as well as scores of military personnel killed and wounded on the ground; however, the attack was not as calamitous as the one at Pearl Harbor, and General Douglas MacArthur, who had been called out of retirement in the summer of 1941 to act as commander of the US Army Forces in the Far East in anticipation of a potential war with Japan, remained in control of the Philippines. There was no rapid capitulation to the Japanese there.

The Philippines would become a protracted campaign, and in that sense, was different than almost every other theater of the war in Southeast Asia. Hong Kong was also attacked on the 8th of December 1941, as Japan effectively went to war with Britain at the same time as the United States, and only held out for two and half weeks. Singapore only managed to hold out for a week against the Japanese in mid-February 1942. The Dutch East Indies were overrun in seven weeks. By way of contrast, the campaign to take the Philippines would last for five grueling months.

When the initial attack on Clark Field occurred on the 8th of December 1941, MacArthur had well over 100,000 men under his command, but the bulk of these were Philippine soldiers, many of them reservists with only minimal training. Additionally, the attack on Clark Field reduced MacArthur's aerial forces, and with the US Pacific Fleet in ruins after Pearl Harbor, the strategic situation was not good for the Philippines, which was soon cut off from any outside support. The Japanese were able to encircle the island of Luzon, where the fighting was concentrated, and establish air superiority, as well. Small contingents of Japanese troops began landing around the large island in mid-December, but it was the 22nd of December before the bulk of the 120,000+ Japanese troops committed to the invasion began landing on Luzon.

Realizing that without air support or outside aid his position was strategically difficult, MacArthur began withdrawing his men onto the Bataan Peninsula on the western side of the Bay of Manila in the final days of December 1941 and beginning of January 1942. This offered the best strategic site to try and hold out against attacks until reinforcements arrived. Over the next three months, the Philippine campaign effectively centered on the Battle of Bataan, where MacArthur still had upwards of 85,000 men and was able to hold his ground against the Japanese through weeks of fighting. However, the battle was more like a siege, as the Americans and Filipinos were fighting against the clock as all of their rations ran low. By March 1942, they were down to 2000 calories of food per day, a fine amount for a modern office worker but entirely insufficient for a soldier operating in a warzone, where 3500 to 4000 calories per day are needed. As rations ran ever lower, it became apparent that surrender was inevitable unless outside support arrived, and there was little hope of that happening.

By mid-March, MacArthur had decided on flight rather than capture. On the 11th of March, he, along with his family and some other senior staff members, broke through the Japanese blockade around Luzon in four PT-20 motorboats, eventually making it to the island of Mindanao in the south of the Philippines and flying from there to Australia. The troops at Bataan held out for another four weeks, only surrendering on the 9th of April to Homma's Japanese forces. What followed was just one example of the huge number of atrocities committed by the Japanese across the Far East during the Second World War: 75,000 Philippine and American POWs were taken on a forced death march from Bataan to Camp O'Donnell, with 500 or so American soldiers and anywhere between 5,000 and 20,000 Filipinos losing their lives in the process. The numbers are widely disputed.

The defeat at Bataan did not bring the Philippines campaign to an end, although it did make the result inevitable. An American contingent continued to hold out for another month on the island of Corregidor, off the coast of Luzon and south of the Bataan Peninsula in Manila Bay. This small island was heavily fortified with gun emplacements, forts, and other defensive structures, and, with its position guarding the waters leading to the city of Manila, had become known as the "Gibraltar of the East." The US military command had been established there after the initial invasion, and it was from Corregidor that MacArthur took flight on his motorboats in mid-March. After the fall of Bataan, the Japanese siege and bombardment of the island, where some 10,000 American and Filipino military personnel were still stationed, intensified. Over the next several weeks, the American gun emplacements were gradually bombed and shelled out of action, allowing for an eventual mass troop landing by the Japanese on the 5th of May. The following day, facing tens of thousands of Japanese troops, the Americans surrendered, bringing the Philippine campaign to an end five months after it began.

The conclusion of the Philippines campaign brought the entirety of Southeast Asia under Japanese rule, giving the Tokyo government control of the oilfields that had been one of its primary targets. The island archipelago would remain under Japanese rule for the next three years. It was, however, a short-lived success. Just a month after the fall of Corregidor and the end of the Philippines campaign, the Battle of Midway took place in the middle of the Pacific Ocean. A major US victory was scored, and in the months that followed, the Americans and their allies stopped the Japanese advances and began pushing them back, while also coordinating their efforts so that the Japanese soon faced a combined alliance of Americans, Australians, Canadians, British, and Chinese in a huge front that stretched from India in the west, through the East Indies and the Philippines to the south, and to the Solomon Islands and beyond in the Pacific. Before long, the Japanese would begin to realize they had brought disaster upon themselves.

SOURCES AND FURTHER READING:

Costello, John. 2009. *The Pacific War, 1941-1945*. New York: Harper Perennial.

Goodman, Grant K. "The Japanese Occupation of the Philippines: Commonwealth Sustained." *Philippine Studies 36*, no. 1 (1988): 98-104. https://doi.org/First Quarter.

José, Ricardo T. "War and Violence, History and Memory: The Philippine Experience of the Second World War." *Asian Journal of Social Science 29*, no. 3, Special Contestations of Memory in Southeast Asia (2001): 457-470.

Pike, Francis. 2016. *Hirohito's War: The Pacific War, 1941-1945*. London: Bloomsbury Academic.

Toland, John. 2003. *The Rising Sun: The Decline and Fall of the Japanese Empire, 1936-1945*. New York: Modern Library.

The Battle of the Kerch Peninsula
(26 December 1941 – 19 May 1942)

It is easy to forget the sheer scale of the war on the Eastern Front during the Second World War, and battles there that eclipsed most of what occurred in Western Europe have been all but forgotten. A good example is the Battle of the Kerch Peninsula. This was a bitter, five-month struggle on the shores of the Black Sea in which slightly less than a quarter of a million Germans and Romanians fought well over half a million Soviets, with Russian defeat clearing the way for the German advance on Stalingrad in the late summer.

Capturing Moscow and Leningrad might have been the two main objectives of the Nazis in the autumn and winter of 1941, but there were other military campaigns taking place within the Soviet Union. Further south, after the capture of Kyiv in mid-September, the main focus of military activity in Ukraine and east towards southwestern Russia was the Crimean Peninsula. The Germans wanted to control this area for numerous reasons. It was a logical next point in their eastward advance towards Rostov and Stalingrad and the oilfields of the Caucasus region that were considered vital for the German war effort, much like the Japanese designs on the oilfields of Borneo and Burma. Additionally, the Crimean Peninsula would open up access to some of the most significant ports on the Black Sea, while the Romanians, the Nazis' foremost ally in the invasion of the USSR, wanted more concrete control over the region and were to be granted additional toeholds in western Ukraine and the western Black Sea once the war was over. Finally, in the deranged plans drawn up by the Nazi leadership for the post-war reconstruction of Eastern Europe, the Crimean Peninsula was to become a sort of tourist resort on the Black Sea for the German upper class.

The Crimea campaign began in October 1941 when Field Marshal Gerd von Runstedt, the overall commander of Army Group South, ordered the German 11th Army under General Eugen von Schobert to break off from the main force of Army Group South, which was heading from Kyiv towards Rostov, and occupy the Crimean Peninsula. The 11th Army, consisting of some 200,000 men, hived off from Army Group South and headed towards the Black Sea. Von Schobert would not command the 11th Army in the Crimea, though. The plane that delivered him to the front there landed in a freshly laid Soviet minefield on the 12th of September, and von Schobert was killed. A much more capable commander, Erich von Manstein, who had largely devised the strategy that led to German victory in the Battle of France in the early summer of 1940, was then appointed head of the 11th Army. He arrived in October, and much of the Crimea was overrun in the weeks that followed by von Manstein's 11th Army and the Romanian Mountain Corps commanded by Lieutenant Gheorghe Avramescu.

The quarter of a million or so German and Romanian forces faced a large contingent of Soviet troops to begin with, but it is important in any assessment of clashes between the Germans and Soviets on the Eastern Front to remember how outmatched the Russians were, tactically and strategically, during the war. Early in 1941 and 1942, between eight and ten Soviet soldiers were killed for every one German. This ratio improved to around six or seven to one as the war went on, but by all standards, the Soviets paid a heavy price in any engagement with the Germans. Thus, while the Crimean Peninsula was theoretically well-defended in October and November 1941, the Germans nevertheless quickly established control over most of it. The key holdout by December 1941 was the port city of Sevastopol in the southwest of the peninsula.

When the Germans laid siege to Sevastopol in late October 1941, the Soviet high command determined a novel approach to the campaign in the Crimea. They would initiate a large-scale amphibious landing on the Kerch Peninsula on the far east of the Crimean Peninsula. The Kerch Peninsula is an important strip of land that controls the access points from the Black Sea into the Sea of Azov, and which can be reinforced through the Taman Peninsula and the Krasnodar region to the east. It was there that the foremost battle of the Crimean Campaign would be fought between December 1941 and May 1942. The goal of the campaign was for the Soviets to secure control of the Peninsula and then apply pressure on the German and Romanian lines to pull their forces away from the siege of Sevastopol, allowing the city to hold out for considerably longer than it otherwise would be able to.

The amphibious campaign to secure the Kerch Peninsula, and to keep reinforcing it, was unusual for the Russians and had not been previously attempted by the Red Army. After some alterations, the operation ended up under the command of Dmitry Timofeyevich Kozlov. In total, between late December 1941 and the early summer of 1942, approximately half a million Soviet troops were transported across the Kerch Strait and landed on the Kerch Peninsula. They faced a numerically inferior German and Romanian force, but as we have seen, numbers counted for little when it came to clashes on the Eastern Front, and the superior tactics, weaponry, command, and use of tanks, artillery, and airpower by the Germans would prove critical.

The Battle of the Kerch Peninsula began on the 26th of December 1941 as troop carriers began landing on the Kerch beachheads. This was a chaotic operation, often carried out using small boats and fishing vessels, with some of the landing sites too shallow to allow an effective or speedy landing. Dozens, or even hundreds, drowned, and over the first weeks of the expedition, thousands of Soviet soldiers developed hypothermia from being dropped into the water and forced to wade ashore in freezing temperatures. Despite this inauspicious beginning, brute delivery of manpower to the Kerch Peninsula ensured that the Soviets established a strong position there by January 1942, reinforced eventually by Russian T-26 Tanks, artillery guns, and anti-tank weaponry.

The Soviets were fortunate that the Germans were unusually slow in responding to the Russian deployment at Kerch due to von Manstein being reluctant at first to pull his troops away from Sevastopol. Because of this, it was mid-January before a major German attack on the Soviet positions on the Kerch Peninsula began. The engagements between the 15th and 20th of January set the tone for the entire battle, as the German 30th Corp clashed with the Russian 44th Army and inflicted thousands of casualties on the Russians, many of which resulted from German Stuka dive-bomber attacks. After several days of intense fighting, nearly 7,000 Soviet soldiers of the 44th were dead, 10,000 had been captured, and scores of tanks and artillery guns had been destroyed, compared to approximately 1,000 dead and

THE BATTLE OF THE KERCH PENINSULA

wounded Germans. A similar rate of attrition continued around the Peninsula for the next three and a half months.

Throughout the spring of 1942, von Manstein continued to make requests back to the German high command, which by then was located at the Wolf's Lair, Hitler's impregnable military bunker in East Prussia, Poland. In these requests, von Manstein called for additional air support in order to displace the Soviets from the Kerch Peninsula. This would, he stated, free up his troops and allow him to break the deadlock at Sevastopol, delivering the entire Crimean Peninsula into German control. Von Manstein's requests, though, were only half-heartedly agreed to, as most of the air support that was available was sent further north to Leningrad, Moscow, and Rostov. In some of the first signs of how thinly German resources were being spread, von Manstein's requests received only limited air support in the form of Stuka dive-bombers.

That all changed in early May 1942, and would lead to swift victory in the Battle of the Kerch Peninsula. In the first days of May, *Fliegerkorps VIII*, the 8th Air Corps of the Luftwaffe under the command of Wolfram Freiherr von Richthofen, a cousin of Manfred von Richthofen, the famous Red Baron flying ace of the First World War, was dispatched to the Crimea. The 8th Air Corps was the most effective close air support unit of the entire Luftwaffe, and in mid-May, worked alongside the German land forces as they initiated Operation Trappenjagd to finally displace the Soviets from the Kerch Peninsula and achieve victory in the Crimean campaign.

The bitter fighting that followed between the 8th and 19th of May could be characterized as a drawn-out bloodbath, particularly once the Russian anti-aircraft guns were largely knocked out of action around the 9th, 10th, and 11th of May. Thereafter, von Manstein ordered von Richthofen's Henkel bomber planes to fly missions over the peninsula dropping anti-personnel bombs. The Henkel bombers were large, slow-moving planes, but they were now largely unchallenged in the skies and inflicted enormous casualties on the Soviet lines. As a result, the Germans began advancing rapidly. The town of Kerch, lying on the northern end of the peninsula, fell on the 15th of May, and the final pockets of resistance were defeated over the next four days. With this done, von Manstein was finally able to commit his full resources to the siege of Sevastopol, and now had the aid of the 8th Air Corps, as well, which dropped 20,000 tons of bombs on the port city in June 1942. Sevastopol capitulated on the 4th of July and tens of thousands of Soviet troops were captured. The Crimean campaign was finally won for the Germans, though it had taken an inordinately long time owing to the diversion created by the Battle of the Kerch Peninsula.

The Battle of the Kerch Peninsula was arguably the costliest engagement of the entire war for the Soviets. The ratio of Soviet to German soldiers lost there exceeded thirteen to one; the Germans and their Romanian allies lost about 40,000 troops, both killed and wounded, while the Soviet casualties are believed to have exceeded half a million men killed, wounded, and captured. There are many reasons for this disparity. The amphibious landings on the Kerch Peninsula in late December 1941 and early January 1942 were botched, with many men killed or wracked by hypothermia afterward. The Germans had air superiority for the vast majority of the five-month-long battle, and the Soviets were often sitting ducks for bomber planes.

Finally, tens of thousands of Soviet soldiers were not successfully evacuated in May 1942, and were instead abandoned by their government on the Kerch beachheads.

Given the sheer brutality of the Battle of the Kerch Peninsula, one would think that it would be better known. After all, the losses there exceeded those of the Western Allies in Normandy and northern France in the two and a half months between D-Day and the liberation of Paris in the summer and early autumn of 1944. And yet the engagement on the eastern side of the Crimean Peninsula is all but forgotten today, at least in the western historiography. It is eclipsed by the Battle of Stalingrad to the east of the Crimea months later, where nearly two million men would be killed, wounded, or captured in what became the crucible of the war on the Eastern Front.

SOURCES AND FURTHER READING:

Burleigh, Michael. 2001. *The Third Reich: A New History*. New York: Hill and Wang.

Forczyk, Robert. 2008. *Sevastopol 1942: Von Manstein's Triumph*. Oxford: Osprey.

Shepherd, Ben. 2016. *Hitler's Soldiers: The German Army in the Third Reich*. New Haven: Yale University Press.

Sweeting, C.G. 2004. *Blood and Iron: The German Conquest of Sevastopol*. Lincoln, NE: Potomac Books.

Tucker-Jones, Anthony. 2016. *Images of War: The Battle for the Crimea, 1941-1944*. South Yorkshire, UK: Pen and Sword Books.

The Aleutian Islands Campaign
(June 1942 – August 1943)

It is seldom mentioned in popular accounts of the war, but the Aleutian Islands, a chain of dozens of small islands extending from Alaska westwards into the Bering Straits, were occupied by the Japanese in the summer of 1942, and remained under Japanese control for over a year before being liberated. The Aleutian Islands Campaign was more important than it is usually credited as having been, as the Japanese could have used the islands to launch bombing campaigns against American cities like Seattle and San Francisco.

In the months following the Japanese attack on Pearl Harbor, the Japanese had virtually everything their way in the war in the Far East and Pacific theater. The British were in no position to commit sufficient resources to try to prevent the fall of Hong Kong and Singapore and further encroachments into Burma, while the US Pacific Fleet had been virtually disabled in Hawaii. Furthermore, President Roosevelt's administration promised the Americans' new allies, the British and the Soviets, that they would focus on defeating the Germans and Italians first, before the Japanese. As such, the resources that could be gathered in America as the country's economy transformed onto a wartime footing were earmarked for a campaign in North Africa, which would eventually become Operation Torch. In this environment, the Japanese empire expanded enormously in the first half of 1942, with the Dutch East Indies added to its territories, Darwin in Australia bombed, and many smaller island chains in the western Pacific occupied.

It was in this context that the Japanese military command decided to invade the Aleutian Islands. The Aleutian Islands are a chain of hundreds of islands and islets extending in a long arc from southwestern Alaska across the Northern Pacific Ocean towards the Kamchatka Peninsula in northeastern Russia. The islands were formed many millennia ago through volcanic activity, as the North American Plate meets the Kula Plate there. A great proportion of the islands are very small and have been largely uninhabited throughout history, both on account of their size, the impossibility of agriculture or any kind of self-sustaining economic activity, and the extremely inhospitable climate. However, there are over fifty smaller islands that can conceivably support some form of life, and fourteen larger islands, such as Adak, Attu, and Kiska, that have sustained communities at one point or another. Adak Island, for instance, still has several hundred inhabitants today, and at one point during the Cold War, had a naval base where thousands of people lived, and even a McDonald's. The largest settlement in the island chain is on Amaknak Island, where over 4,000 people live in the town of Unalaska, but this is the exception to the generally uninhabited nature of the island archipelago. It is not hard to see why it is so thinly inhabited. Given the northerly location, the islands experience very little sunlight for several months of the year, and temperatures are generally below 20 degrees Fahrenheit (-7 degrees Celsius) in winter. On Kiska, climate studies recorded an average of 150 days of snow per year in the middle of the twentieth century.

It was this chain of islands that the Japanese decided to occupy in the summer of 1942 as part of their military expansion. The Aleutians generally are part of Alaska, and it was believed by the government in Tokyo that they could be used as a springboard to invade mainland Alaska in due course. Furthermore, if runways and airfields could be established, the Aleutian islands might be used as a location, close enough to the West Coast of the United States, from which Japanese planes could bomb cities like Seattle and San Francisco, which was not possible from any other Japanese territory as bomber planes could not travel that far in the early 1940s. Finally, the occupation of the Aleutians would serve a defensive purpose, too. Just as the Japanese could conceivably use the islands to launch bombing missions against American cities, they believed that the Americans could also use them to initiate aerial attacks against the Home Islands of Japan. The emperor and his generals were particularly concerned about this possibility after the Doolittle Raid, in which the US launched attacks against Tokyo and other parts of Japan on the 18th of April 1942.

The Aleutian Islands were not entirely uninhabited at the time. There were two naval and military bases on the islands, one at Fort Glenn Army Airfield on Umnak Island, the other at a site called Dutch Harbor on Amaknak Island. A third military garrison was stationed at Cold Bay on the mainland of Alaska, near the Aleutian Islands. In total, there were over 2,300 American soldiers and military personnel across these three military installations, but they were easily overcome by the Japanese in early June 1942 when an expedition led by Admiral Kakuji Kakuta and consisting of troop carriers with approximately 8,000 Japanese soldiers, several destroyer-class battleships, three cruisers, and some submarines, arrived at the Aleutians. They were accompanied by two small Japanese aircraft carriers with a mix of A6M Zero fighters and both torpedo and dive bombers.

The first major engagement of the campaign took place on the 3rd of June 1942, when the Japanese attacked the US military base at Dutch Harbor on Amaknak Island. There was a considerable exchange of fire, and the Japanese launched planes against the base. After the initial engagements, the Japanese decided to focus their efforts on some of the other islands, landing on Attu and Kiska between the 5th and 7th of June. These would become the focus of their occupation over the next year. Thus, with a limited campaign, the Japanese had claimed control over much of the Aleutians and begun bunkering down on several islands. The attacks on the Aleutians gave the Japanese a potential additional advantage, as well: they could pull US forces northwards in the days leading up to the Battle of Midway between the 4th and 7th of June 1942. This initiative, however, failed.

Back in the United States, public sentiment was alarmed when news of the Japanese occupation of the Aleutian Islands was received. The US is a country that, much like Britain, has rarely experienced attacks on its own sovereign territory in modern times; one has to go back to the War of 1812 to find enemy troops actually occupying American territory. Pearl Harbor had been a hit-and-run mission. The Aleutians, however, would be controlled by the Japanese for a year. Despite the outcry over the occupation, the US military was in little position to do anything about it in the short term. Preparing an expedition to head north to Alaska would take time, and it would not be possible to mount a major offensive in the Alaskan winter. Thus, it was decided early on that the islands would not be completely liberated until 1943. The US Navy did keep the Japanese on their toes, though. In early July 1942, one of their

THE ALEUTIAN ISLANDS CAMPAIGN

submarines, the USS *Growler*, embarked on a mission to Kiska in which it managed to fire off a barrage of torpedoes that sank a Japanese destroyer ship and damaged two more before escaping unharmed. Occasional bombing missions were also launched from Alaska, but the Japanese had spread out over Attu and Kiska and did not present easy targets for the US Air Force. The silver lining was that no Japanese activity to begin establishing extensive air installations on the islands, or utilize them as bases for further attacks on the US mainland, were detected.

While there were sporadic clashes between the Japanese and the Americans in and around the Aleutians in late 1942 and early 1943, the real campaign to retake the islands did not commence until mid-May 1943, when Operation Landcrab was launched after securing the sea lanes in the Northern Pacific in March and April. An initial wave of over 11,000 US and Canadian troops landed on Attu. By this time, approximately 2,500 Japanese were stationed there under the command of Colonel Yasuyo Yamasaki, with the remainder of the expeditionary force largely on Kiska.

The Battle of Attu raged between the 11th and 29th of May and was the main engagement of the entire Aleutian Islands campaign. Over nearly three weeks of intense fighting, the Americans and Canadians experienced brutal weather conditions, zealous Japanese opposition, and an environment in which booby traps had been placed all around the island in preparation for an Allied invasion. Yamasaki had also prepared well for the engagement, moving his men to the high ground and establishing defensive points at intervals around the island. Moreover, the Japanese generally fought to the death at every juncture, and even the last engagement on the 29th of May climaxed in a *banzai* charge, wherein the Japanese refused to be taken captive. In the end, over 90 percent of the Japanese were killed, with only a few dozen taken captive. The Japanese medics even killed their own wounded by throwing grenades into the field hospital before the final banzai charge. The US and Canadians suffered over 500 casualties, while thousands more were wounded liberating the island.

The Americans and Canadians began preparing to liberate Kiska and end the Aleutian campaign in the days that followed, expecting to meet a larger number of Japanese troops and even firmer resistance when they landed there in mid-August. Consequently, it was a larger Allied force of over 30,000 troops that were sent to the island. However, they were not needed. When they arrived in mid-August, they discovered that the Japanese had abandoned the island a few weeks earlier, at the very end of July. US intelligence had failed to detect this, as the withdrawal was undertaken during a bout of intense fog that fell over the island chain late that summer and cloaked the Japanese departure. Thus, the Aleutian Islands campaign ended in something of an anti-climax, but the Battle of Attu in May 1943 had seen intense fighting on American territory, the only land combat on American soil during the Second World War.

That the Japanese ultimately elected to abandon Kiska is unsurprising. By the late summer of 1942, there were already distinct signs that the war effort was turning against the Japanese in the Pacific. First, the US victory at Midway had swung the naval campaign in the Pacific in favor of the Americans, while the Guadalcanal campaign was commencing

THE ALEUTIAN ISLANDS CAMPAIGN

in the Solomon Islands even as troops were searching Kiska to confirm that the Japanese had abandoned the island. The US would briefly consider utilizing the Aleutians as an aerial base to strike at the Japanese, but the weather conditions and the difficulties of trying to establish airfields there in a short period of time prevented them from doing so, much as they had for the Japanese during the year they occupied the islands. In the end, the major enemy faced by both the Japanese and the Americans was the weather.

SOURCES AND FURTHER READING:

History.com Editors, "Battle of the Aleutian Islands," History.com, November 17 2009, https://www.history.com/topics/world-war-ii/battle-of-the-aleutian-islands.

Garfield, Brian . 1995. *Thousand-Mile War: World War II in Alaska and the Aleutians*. Fairbanks, AK: University of Alaska Press.

Harris, Megan. "The Aleutian Islands: WWII's Unknown Campaign." Library of Congress Blogs. March 14, 2014. https://blogs.loc.gov/loc/.

Perras, Galen R. 2003. *Stepping Stones to Nowhere: The Aleutian Islands, Alaska and American Military Strategy, 1867-1945*. Annapolis, MD: Naval Institute Press.

Roblin, Sebastian. "Forgotten World War II Fact: Japan 'Invaded' Alaska." *The National Interest* (Washington, D.C.), February 21, 2020.

The Battle of Algiers
(8 November 1942)

The Battle of Algiers was one element of Operation Torch, the wider campaign that saw the Americans arrive in North Africa to aid the British against the Italians and Afrika Korps. The battle was fought between the Western Allies and the French pro-Vichy administration of Algeria. The Allied attack on Algiers and the capitulation of the city was more significant than the limited military action would suggest, as it led to the military occupation of Vichy France by the Nazis and the scuttling of the French Mediterranean fleet in retaliation.

The Japanese attack on the United States in December 1941 did not just create a war for the US in the Pacific and the Far East. Owing to Japan's alliance with Germany, America was also at war with the Axis Powers in Europe, an eventuality Winston Churchill had been encouraging for two years. President Roosevelt had also wished to end neutrality and enter the European war for some time, but public sentiment in America was isolationist until Pearl Harbor. Just over two weeks after Pearl Harbor, Roosevelt and his government welcomed British delegates to Washington D.C. for the Arcadia Conference. At the US capital for Christmas 1941, Churchill and Roosevelt came to an agreement that the Allied nations would commit to defeating Germany and its allies in Europe first, then focus their attentions on the Empire of Japan and the Pacific theater. The German invasion of the Soviet Union was still at a critical juncture in Russia, but Stalin and his officials were most certainly in favor of the approach decided upon in Washington.

There was a complication, though. The Western Allies lacked a bridgehead on the European continent from which they could strike at the Germans. The Nazis controlled everything from the Pyrenees along the border of Nazi-occupied France and General Francisco Franco's Spain, a country which was friendly to Germany but which had demurred from entering the war, all the way north to northern Norway inside the Arctic Circle. Extensive defenses known as the Atlantic Wall were being built all along this massive coastal perimeter to fend off any Allied efforts to invade Western Europe. Furthermore, the US, British, and their allies, such as the Commonwealth powers of Canada and Australia, still had only limited resources available as 1942 dawned. It would take many months for men to be conscripted and trained in the US and for the American economy to begin producing Sherman tanks instead of Ford automobiles. Roosevelt and Churchill decided at the Arcadia Conference that they would focus in the meantime on the soft underbelly of the Axis war effort in North Africa.

As we have seen, after the spectacular British victories at Beda Fomm and elsewhere in Libya in late 1940 and early 1941, the German Afrika Korps had been sent to North Africa. Under Erwin Rommel, Afrika Korps had achieved its own major victories along with the Italians, pushing the British and Commonwealth divisions back into eastern Libya, and then into northwestern Egypt. The Western Desert Campaign, as this to-and-fro across the Saharan Desert west of the Nile River became known, consisted of the two sides pushing each other over and back across the Libyan and Egyptian border for many months in the second half of 1941 and the first half of 1942, with major clashes occurring at the Siege of Tobruk and other sites. By the time Churchill left Washington in January 1942, the Germans and Italians were launching a fresh counterattack into western Egypt, but British and Commonwealth victories in the summer and autumn of 1942 at the First and Second Battles of El Alamein, just over a hundred kilometers to the west of the city of Alexandria in northern Egypt, prevented the Axis from advancing towards the Nile and threatening Cairo or the Suez Canal. Nevertheless, by the time the new Allied campaign against North Africa was launched, the Italians and the Germans still controlled most of Libya and Tunisia, while the French Vichy collaborationist government controlled Algeria and Morocco, leaving the Axis Powers in control of the vast majority of North Africa.

While Roosevelt and Churchill had agreed in principle at the Arcadia Conference to begin a major Allied offensive in North Africa, the details of exactly what this would involve would take months to be decided, and would evolve with the changing strategic situation on the ground in the Western Desert. The plan that was eventually decided upon was codenamed Operation Torch. An amphibious campaign would be launched with American, British, Canadian, Commonwealth, and Free French and Dutch soldiers landing in several locations along the coast of Morocco and Algeria to secure the colonies controlled by the Vichy France there before proceeding east toward Tunisia. At the same time, a reinforced British army would continue to press into Libya from Egypt, attacking the Italians and Germans from both sides and hopefully securing all of North Africa before long. The plan would have the added advantage of being a trial run for launching an amphibious invasion of Western Europe at some later date.

Torch was planned for November 1942. General Dwight Eisenhower was placed in overall command of the 100,000 troops involved, the greater proportion of which would be provided by the US. Eisenhower was seconded by General George S. Patton, arguably the finest military commander on the US side during the whole of the war, though a difficult individual who courted controversy and was something of a loose cannon. There were to be three distinct task forces, with three distinct targets, within Operation Torch. The Western Task Force, led by Patton, was prepared in Virginia and sailed directly from there for Morocco, where its main targets were Casablanca, Mehdia, and Safi. The Central and Eastern Task Forces were gathered in Britain and sailed from there, with their targets being the cities of Oran and Algiers on the north coast of Algeria. The Eastern Task Force, headed for Algiers, was commanded by Major General Charles W. Ryder, an American general, although of all the task forces, the Eastern Force had the majority of British troops. Ryder had overall command of just over 30,000 men, approximately two-thirds British and one-third American. The Eastern Task Force planned to arrive at Algiers on the 8th of November 1942, after sailing through the Straits of Gibraltar, where control of the Rock by the British assured safe Allied passage into the Mediterranean.

The Battle of Algiers commenced on the 8th of November as thousands of Allied troops began disembarking from their ships onto beachheads to the west and east of the city of Algiers. It must be said that as far as trial runs for D-Day went, the Operation Torch landings around Algiers were broadly pointless. The Western Allies simply arrived at landing locations and got off the ships without facing any real resistance, although stiffer combat conditions were experienced in Morocco and Oman. Once on land near Algiers, the divisions began to move towards the city. As they did, they learned why there had been so little initial resistance to their landings. The previous day, a coup of sorts had occurred in Algiers, as the Vichy administration was overthrown by a faction that was more amenable to negotiating the surrender of the city to the Allies. Still, street-to-street fighting began as the Allied troops entered Algiers, even as these negotiations were underway.

The negotiations soon began to center on one man: Admiral Francois Darlan. Darlan was a decorated naval commander who had been serving as commander-in-chief of the French navy when the Second World War began. However, he subsequently disgraced himself by collaborating as part of the Vichy regime. He was appointed to several Vichy ministerial briefs, including Minister of the Interior and Vice-President of the Council, effectively becoming the second most senior figure in the regime next to Philippe Pétain. In April 1942, though, Darlan had been forced to step down from most of his offices by the Germans, who didn't trust his intentions. Perhaps they were right, for within hours of the beginning of the Battle of Algiers, Darlan was in contact with Eisenhower about relinquishing control of the city to the Allies and commanding the Vichy French forces in North Africa to switch sides in the war.

Eisenhower quickly accepted Darlan's offer, much to the dismay of the Free French leader in London, Charles de Gaulle, and many British politicians. The exact circumstances around Darlan relinquishing control remain unclear. Some contend that Darlan was in contact with the Allies for several months and that his presence in Algiers when Operation Torch began was more than a mere coincidence. In any event, not long afterward, the city was surrendered to the Allied forces. Furthermore, the port was secured intact. A special operatives mission codenamed Operation Terminal was launched before dawn that morning to land 600 troops in the port of Algiers, where their goal was to ensure that the Vichy French didn't attempt to destroy the port facilities before the city fell and potentially deprive the US and British of its use for weeks to come. By sundown on the 8th, the city was in Allied hands and the port was still fully functional, allowing men and materiel to flow into this part of North Africa in the weeks to come.

Incredibly, it was soon revealed that Eisenhower's deal with Darlan included a provision whereby Darlan became the High Commissioner of the Free French in Africa and effectively the new French governor of Algeria and Morocco. This had the added benefit of Darlan being able to convince some of the Vichy French in Senegal and the other French colonies of West Africa to switch sides to support the Allies. Yet de Gaulle and others were appalled by this collaboration with an individual who had been so high up in the Vichy administration. The level of resentment towards Darlan was so high that he did not live long enough to serve as High Commissioner. He was assassinated in his offices in Algiers on the 24th of December 1942 by Fernand Bonnier de la Chapelle, an anti-Vichy Frenchman.

The Battle of Algiers, in and of itself, might have been a relatively inconsequential military clash, but its implications were wide-ranging. In Western Europe, when the Germans learned of Darlan's betrayal, they moved to occupy much of the

THE BATTLE OF ALGIERS

territory of Vichy France, in breach of the agreement that the Vichy-controlled parts of the country would not be occupied by German troops. The Germans were particularly anxious to secure the Mediterranean coastline of France, which was now exposed to potential Allied attacks from Algeria. They also planned to commandeer the French fleet stationed at the port of Toulon, in a clear breach of the armistice agreement of 1940. This was too much for the Vichy regime, and Pétain and his associates ordered the scuttling of the entire fleet at Toulon on the 27th of November before it could fall into German hands. When British RAF reconnaissance planes flew over southern France the following day, they captured photos of dozens of French warships and cruisers either ablaze or sinking in the harbor of Toulon. Hence the Battle of Algiers had knock-on effects in France that led to the occupation of Vichy France's territory by the Germans, a serious sundering of relations between the Nazis and Pétain's administration, and the destruction of the French fleet.

More broadly, the success of Operation Torch across Algeria and Morocco led the Allies to prepare to launch an attack on Tunisia, which was more stoutly defended by the Italians and Germans, early in 1943. By that time, the British and their Commonwealth allies had conclusively gained the upper hand in Libya and were pressing the Axis divisions there towards southern Tunisia. The Tunisian Campaign would see intense fighting through the spring of 1943, as by then the Axis position had been heavily reinforced by the Germans. Only in mid-May 1943 was the campaign victorious, leaving the Allies in control of North Africa from Casablanca to Cairo. Nearly two months later, in early July 1943, Operation Husky, the Allied invasion of Sicily and the opening of a Southern Front in Italy, would begin.

SOURCES AND FURTHER READING:

Anderson, Charles R. 2015. *Algeria-French Morocco: The US Army Campaigns of World War II*. Scotts Valley, CA: Create Space Independent Publishing.

Breuer, William B. 1985. *Operation Torch: The Allied Gamble to Invade North Africa*. New York: St. Martin's Press.

Funk, Arthur L. "Negotiating the Deal with Darlan." *Journal of Contemporary History 8*, no. 2 (1973): 81-117.

O'Hara, Vincent P. 2015. *Torch: North Africa and the Allied Path to Victory*. Annapolis, MD: Naval Institute Press.

Trouillard, Stephanie, and Tom Wheeldon. "Allies' Successful First Invasion but a 'Botched' Job: Operation Torch, 80 Years On." France24.Com. November 8, 2022. https://www.france24.com/en/africa/20221108-allies-successful-first-invasion-but-a-botched-job-operation-torch-80-years-on.

Operation Gunnerside
(February 1943)

> The United States was not alone in trying to develop a nuclear weapon during the Second World War. The German equivalent of the Manhattan Project was the "heavy water" project carried out in southern Norway. The Allies knew what the Germans were up to, and in February 1943, launched a special operations mission, codenamed Operation Gunnerside, to destroy the production facility and end the threat of a German nuclear weapon.

The Second World War was not all large battles involving tens or hundreds of thousands of men at a time. Many important objectives were achieved using small bands of special operatives or units of troops trained in specific ways. The Germans, for instance, were able to secure control of many Norwegian cities and towns in April 1941 using small groups of paratroopers and naval divisions, and they took over the Scandinavian country this way. In North Africa in 1941 and 1942, the British SAS was born as crack units of desert operatives commanded by Major David Stirling began striking at German and Italian military installations in the Western Desert. British special operatives launched dozens of missions against German naval bases and intelligence centers in France and the Low Countries between 1940 and 1944, missions like the St Nazaire Raid of March 1942, where a few hundred British commandos managed to destroy the port of St Nazaire in Brittany in France and deprive the Nazis of its use at a critical time in the Battle of the North Atlantic. There were many other important missions like this.

Surely one of the more critical special operations missions in this respect, though one which rarely, if ever, features in histories of the Second World War, was Operation Gunnerside, which paralyzed the Nazi project to develop a nuclear weapon in February 1943. We are all familiar with the Manhattan Project, the US nuclear weapons program during the Second World War that produced the atomic weapons that were dropped on Hiroshima and Nagasaki in August 1945. But it was not just the Americans who were trying to develop nuclear capabilities during the war. The British briefly had their own nuclear research programs underway in several universities and other locations in England in 1940 and 1941; they subsequently abandoned their efforts and began aiding the US program. The Japanese were also carrying out research under the leadership of the physicist Yoshio Nishina, who knew both Albert Einstein and Niels Bohr. Finally, the Germans had their own nuclear program.

Nazi Germany was better positioned than any other nation to produce a weapon of mass destruction when the Second World War began in 1939. Germany had been the epicenter of the scientific world for decades, and ten of the Nobel Laureates in Physics between 1901 and 1939 had come from Germany, while others, like the Dane Niels Bohr, worked in regions that quickly came under Nazi occupation in 1940. If any country was going to develop a nuclear weapon during the war, one would have predicted that Germany would, although Germany's effort was not helped by the flight of some of the greatest minds of the time from Europe to America in the pre-war years or the first year of the war, notably Albert Einstein and Enrico Fermi.

OPERATION GUNNERSIDE

Einstein and others had warned the US government, even before it entered the war, that the Nazis were trying to acquire a nuclear weapon. Ironically, German efforts would focus on research that Fermi had published years earlier. In 1934, the Italian physicist had revealed that it was possible to enrich uranium by overloading it with neutrons, and that this could then produce immense power. The process was inherently unstable, though. What the wartime Nazi scientists determined was that the process of enrichment could be controlled and stabilized if "heavy water"–water in which an isotope called deuterium is substituted for hydrogen–was used to produce heavy water or "heavy hydrogen." One of the benefits of conquering Norway in April 1940 for the Nazis, in addition to its many North Atlantic ports, was that the Vemork hydroelectricity plant in southern Norway in the Telemark region was already producing substantial amounts of heavy water by the end of the 1930s. The Nazi nuclear program would be based out of Vemork after the German conquest of Norway in the late spring of 1940.

The Allies benefited from the fact that the supply of heavy water at the Vemork plant had been removed to France, and from there to Britain, in 1939 in preparation for a potential attack by the Nazis when the war broke out. Consequently, there was no raw material to give the Germans a head start when they began their nuclear program at Telemark in 1940. Furthermore, much like producing enriched uranium for the Manhattan Project in America, the production of heavy water took time, and so it would be several years before the Nazis would ever be able to produce a working bomb. Nevertheless, the British, who knew about the threat posed by the program, along with their growing array of Allies, were determined to sabotage these efforts in Norway. There were two special operations missions to southern Norway before Operation Gunnerside: Operation Grouse and Operation Freshman, which took place in October and November 1942. Grouse's main aim was to place several Norwegian Resistance fighters in the Telemark region in preparation for Operation Freshman weeks later. This second mission saw the landing of a team of British commandos near the Vemork plant. However, Operation Freshman turned into a debacle, with the Germans quickly detecting the British and either killing or capturing all of their operatives. The only advantage of these early efforts was that the Norwegian Resistance fighters who had been airlifted into the Telemark area through Operation Grouse had remained undetected and were able to continue sending intel back to England about developments they observed around Vemork.

Operation Gunnerside resulted in the breakthrough success that the British and Norwegians were hoping for. It was launched on the 16th of February 1943 and involved entirely Norwegian personnel, though using British airplanes and British military support. Six new Norwegian Resistance commandos were airlifted into the Telemark region. They spent some time searching for, and eventually locating, the Norwegians from Operation Grouse, a group that had been codenamed Swallow in the intervening period. Once the operatives from Grouse had established contact with Swallow, they began preparing to strike at the heavy water plant, no easy feat since the Nazis had mined the surrounding area and added additional defenses, including numerous watchtowers, in the aftermath of Operation Freshman.

OPERATION GUNNERSIDE

The full team of eleven Norwegians moved on the night of the 27th of February 1943. The decision to send Norwegian Resistance fighters rested in part on their familiarity with the weather conditions in Norway in the winter and early spring months, and the operatives certainly needed to be inured to the cold. They had to swim through freezing cold water in order to sneak into the hydroelectric plant without being detected by the German guards and defensive points. They managed to gain access to a basement at Vemork and had a good idea of where they were going once inside. Unbeknownst to the Germans, a Norwegian Resistance fighter was working as a secret agent within the plant and had managed to produce schematics of the installation, which had been sent to Britain before the missions were launched in late 1942. Once inside, the Resistance fighters also stumbled upon a Norwegian caretaker who was sympathetic to their mission and gave them further aid.

The commandos were able, with the help of the map and the caretaker, to quickly locate the electrolysis lab. This was where the most critical work in producing heavy water was undertaken and where a large store of the heavy water was located. Setting explosive charges around the lab, the commandos quickly wired these to go off shortly after they left and then made a swift getaway. They were still close enough to the site to hear the explosion when the charges detonated not long afterward, destroying the lab and the supply of heavy water that had taken nearly two years to produce. The Norwegians split up thereafter, believing that they stood a better chance of avoiding detection this way. Four of them remained in the region to join a unit of the local Norwegian Resistance, but others traveled vast distances to make their way out of Norway, some by skiing over the border to neutral Sweden.

Since the Nazis would have needed several tons of heavy water to manufacture a nuclear weapon, they were far from having enough to produce a working device, but nevertheless, Operation Gunnerside completely ended any hopes Hitler and the Nazi high command had of acquiring such a weapon. To put this in perspective, if the Oak Ridge facility in Tennessee, where uranium was being enriched by the Americans in 1943 and 1944, had been sabotaged by an Axis team of special operatives, it almost certainly would have prevented the US from utilizing nuclear weapons against Japan in August 1945. Although it is hard to remember in light of the enormous stockpiles of nuclear weapons developed in the 1950s and 1960s, enriching uranium or producing heavy water or any other raw material that might be used to produce nuclear fission was a time-consuming and hugely expensive process in the early days of nuclear weapons development. Therefore, a mission like Operation Gunnerside was able to completely undermine the Nazi nuclear program.

It was an enormous boon to the people of Europe that the Nazi nuclear weapons program was foiled. Had it continued and a breakthrough been made, no matter how unlikely, it would have spelled devastation for the continent. Hitler and the Nazi leadership would almost certainly have used any such nuclear weapon in their desperation to turn the tide of the war they were losing, unleashing them on cities like London, Moscow, and Leningrad. The Holocaust, the mass use of slave labor in German factories, and the proposed "Hunger Plan"–through which the Germans intended to starve to death tens of millions of people in Eastern Europe if they had won the war against the USSR–indicate the Nazis' pathological inclination towards mass murder. As such, it was to everyone's benefit across the continent that the heavy water program was effectively destroyed in 1943 through Operation Gunnerside.

SOURCES AND FURTHER READING:

Gallagher, Thomas. 2010. *Assault in Norway: Sabotaging the Nazi Nuclear Program.* Guilford, CT: Lyons Press.

Grunden, Walter E., Mark Walker, and Masakatsu Yamazaki. "Wartime Nuclear Weapons Research in Germany and Japan." *Osiris 20: Politics and Science in Wartime,* (2005): 107-130.

Herrington, Ian. "The SIS and SOW in Norway 1940-1945: Conflict or Co-operation?" *War in History 9,* no. 1 (2002): 82-110.

Lobbell, Jarrett A. "The Secrets of Sabotage." *Archaeology* (Long Island City, NY) https://www.archaeology.org/issues/283-1801/trenches/6198-trenches-norway-wwii-heavy-water-plant.

Rønnenberg, Joachim. "Operation "Gunnerside": The Race for Norwegian Heavy Water, 1940-1945." *Norwegian Institute for Defence Studies 4,* no. 95 (1995): 13-16.

Black May
(May 1943)

The Battle of the North Atlantic is a well-known blanket term for the naval war between the British and Germans that took place across the North Atlantic between 1940 and 1944. Yet other than stories about the British cracking the German Enigma code at Bletchley Park, few individual aspects of the wider battle are ever discussed. Surely the events of May 1943, which became known as "Black May" to the Nazis, should be more widely known: over the space of just four weeks, the Nazis lost over forty submarines, a quarter of their operational U-boats, making Black May the defining turning point in the Battle of the North Atlantic.

It is easy to forget that there was fighting in the Second World War that took place far away from the main theaters in Western Europe, Eastern Europe, North Africa, and the Pacific. For instance, there were clashes early on between the British and the Italians in the Horn of Africa. Italy had conquered Abyssinia (modern-day Ethiopia) in the mid-1930s to add it to its colonies in Somaliland, and when Mussolini entered the war in the summer of 1940, this immediately led to clashes between the Italians and the British colonies in Kenya and Uganda. This was not the only out-of-the-way clash of the war, however. Another occurred early on, when the German battleship *Admiral Graf Spree* clashed with British ships near the mouth of the River Plate in South America in December 1939.

Indeed, the war at sea provided some of the foremost examples of engagements in out-of-the-way places. For instance, German submarines, or U-boats, patrolled the waters off the East Coast of the United States, hoping to strike at US shipping, after America entered the war in December 1941. Submarine warfare also led to clashes in some remote regions, with U-boats sinking in Cuba and Barbados, and U-505 captured intact in the waters north of the Cape Verde Islands off the coast of western Africa. Yet much of this submarine warfare is forgotten, and no element of it was more consequential than "Black May," the name given to May 1943, when the Allies turned the tide on the submarine campaign.

Black May was a crucial part of the wider Battle of the North Atlantic, a struggle that has been called the longest campaign of the Second World War. In many ways, the battle for control of the shipping lanes of the North Atlantic lasted the entirety of the war, from September 1939 to May 1945, as clashes occurred between the British Royal Navy and the German Kriegsmarine throughout the period. The most intense period of fighting in the North Atlantic, though, began in the summer of 1940 and followed the rapid successive conquests of Denmark, Norway, the Netherlands, Belgium, and France by the Germans, which gave Germany control of the entire coastline of Europe from the Arctic Circle to the Pyrenees. Thereafter, a major drive of the war at sea began, with the goal of bringing Britain to its knees by cutting the island off from outside supplies. The North Atlantic would continue to be a major theater of the war until the late summer and early autumn of 1944, when the Allied liberation of France and Belgium severely restricted German activities in the Atlantic. Black May was a major component of the Allied victory in the wider battle.

The Battle of the North Atlantic involved several different kinds of clashes, some between large surface battleships, others involving a combination of air and sea power. Yet the element that has captured the imagination, and on which the result hung in the balance between 1940 and 1943, was a game of cat and mouse between the Allied convoys and the German submarine "wolfpacks." In the dark days of 1940 and 1941, in order for the British to continue to bring supplies of war materiel, foodstuffs, and other necessary goods from North America and other regions to their island fortress, they had to get past attacks by German ships and submarines. To do so, the British quickly began organizing large-scale convoys. These convoys involved dozens of large transport ships carrying goods of all kinds–even tanks and armored vehicles–that were supplied to Britain through the Lend-Lease agreement with America before the US entered the war. These flotillas were guarded by as many as a dozen ships of the British Royal Navy, usually a combination of two or three battleships and smaller frigates and cruisers. In response, the Germans began organizing their U-boat submarines into groups of a dozen or more that prowled the North Atlantic looking for convoys to attack from beneath the water. These became known as the wolfpacks.

The Battle of the North Atlantic experienced numerous shifts and turns in the game of cat and mouse that played out between the German wolfpacks and the British convoys from the summer of 1940 to Black May in 1943. When the campaign to pummel the British into surrender after the Battle of France began in the summer of 1940, the Germans had a number of major advantages, chief among them their seemingly undecipherable messaging system, the Enigma code, which allowed them to transmit messages around the North Atlantic and strike British convoys with torpedoes from beneath the waves, then disappear before the British had any time to react, leaving ships sinking in their wake. As a result, the period between June 1940 and February 1941 became known to the Germans as *Die Glückliche Zeit*, "The Happy Time," as they were able to sink Allied ships with enormous success and few German casualties.

Less happy times followed for the Germans, however, as the British began employing better tactics to guard their convoys, including the extensive use of depth charges to strike at submarines underwater. Even more importantly, in the summer of 1941, the Enigma code was deciphered at Bletchley Park in England by code breakers, led by Alan Turing, using a captured Enigma machine. This evened the playing field in the Battle of the North Atlantic. Moreover, the entry of the US into the war later that year increased the Allied firepower along the North Atlantic shipping routes. There were still many U-boats prowling the waters, though, and as the Americans sought to send additional troops to Britain to prepare for operations in Europe, the U-boats were a major problem. Black May would turn the tide firmly in the Allied favor.

Black May began, somewhat counterintuitively, in late April. The Germans still had about 240 functional U-boats, down from a peak of approximately 350 in 1941, but still a sizable naval force. At any one time, only around half of these U-boats were fully operational, with the others docked in ports in France and elsewhere to undergo repairs or allow their crews to go on shore leave while the subs were resupplied. The full force of the Royal Navy and the merchant navy of Britain, with additional support from the US Navy and other allies such as Canada, was arrayed against the German U-boats.

The first major clash surrounded Convoy ONS 5, which consisted of just over forty transport and merchant navy ships carrying tens of thousands of tons of goods bound for Britain. They were accompanied by a detachment of the Mid-Ocean Escort Force, a division of the Royal Navy consisting of seven ships, including the Destroyer-class battleship HMS *Duncan*. The convoy followed one of the main shipping lanes between North America and Britain along the coasts of eastern Canada, Newfoundland, Iceland, and then southwards to Scotland in late April and early May. Most of the engagements took place in the waters off the extensive coast of Newfoundland, where a wolfpack estimated to contain at least thirty U-boats, and perhaps as many as forty at one point, prowled. A week of intermittent attacks followed, with the British able to track some of the submarines–thanks to new techniques using radar–and then strike at them with depth charges and other weapons. Consequently, whereas back in The Happy Time the Germans would have destroyed much of the convoy with only limited losses themselves, over the space of a week, the wolfpack lost at least seven U-boats, with a similar number hit and damaged. The British lost thirteen of their ships, most of them merchant navy vessels. What this amounted to was a Pyrrhic victory, a military clash that is technically a victory, but which comes at such a high cost that it must be viewed as a defeat in many ways. The Germans had sunk thirteen ships, but at this stage in the Battle of the North Atlantic, losing seven U-boats and having another seven forced back to port was far too costly a result for the Nazis.

The attack on Convoy ONS 5 foreshadowed the wider events of Black May. Over the course of the weeks that followed, there were several wolfpack attacks on British convoys, but all experienced similar or worse results than what had occurred with Convoy ONS 5. Some convoy attacks even resulted in more U-boats sunk than British ships. Thus, by the end of May, in excess of forty U-boats had been sunk, while three dozen more were damaged and had to be sent back to France and other European ports for extensive repairs. This constituted one-third of the entire U-boat fleet, over half of those that had been fully operational in mid-April, while only fifty-eight Allied ships in total were sunk during the same period. The U-boat fleet would never fully recover from this major loss of capability in May 1943.

The impact of Black May on the Battle of the North Atlantic and German submarine activity was immense. Initially, all U-boats were ordered back to German-held ports while the head of the Kriegsmarine, Admiral Karl Donitz, who later served as president of Germany for a month after the suicides of both Hitler and Joseph Goebbels, began to work out a new strategy with the very limited number of U-boats that were left at his disposal. With the Soviets now on the offensive in Russia and the Western Allies expected to open a southern front somewhere in the Mediterranean later that year, there were simply no resources available back in Germany to rebuild the U-boat fleet back to anything like what it had been prior to Black May. As a result, from June 1943 onwards, the U-boats would be used for more specific missions and calculated attacks, and efforts to restrict Allied shipping in the North Atlantic by widely deploying U-boats came to an end.

One peculiar and largely forgotten by-product of Black May was the formation of *Monsun Gruppe*. "Monsoon Group" was a detachment of U-boats that were dispatched in the summer of 1943 to sail all the way from Europe to the South Atlantic, around Africa and to Southeast Asia. There they used Japanese ports in Malaysia and other locations to resupply before beginning missions to attack Allied shipping in the Indian Ocean. The primary targets were British ships sailing from British India back to Europe. Nearly twenty U-boats served in the Indian Ocean theater as part of Monsun Gruppe in the second half of 1943 and into 1944, sinking many Allied ships. Nevertheless, despite the limited success of this campaign, the fact that the Germans largely gave up on any efforts to restrict Allied shipping in the North Atlantic in favor of striking at less protected targets in the Indian Ocean was a sign of how desperate the war at sea had become. Hence, Black May was a quasi-conclusive episode in the Battle of the North Atlantic.

SOURCES AND FURTHER READING:

Assmann, Kurt. "Why U-Boat Warfare Failed." *Foreign Affairs 28*, no. 4 (1950): 659-670.

Bruning, John. 2017. *Battle for the North Atlantic: The Strategic Naval Campaign that Won the War in Europe.* Sarasota, FL: Crestline Publishing.

Faulkner, Marcus, and Christopher M. Bell. 2015. *Decision in the Atlantic: The Allies and the Longest Campaign of the Second World War.* Lexington, KY: Andarta Books.

Paterson, Lawrence. 2004. *Hitler's Grey Wolves: U-Boats in the Indian Ocean.* Berkshire, UK: Greenhill Books.

Schofield, B.B. "The Defeat of the U-Boats during World War II." *Journal of Contemporary History 16*, no. 1 (1981): 119-129.

The Battle of the Dnieper
(August – December 1943)

The sieges of Moscow and Leningrad and the battles of Stalingrad and Kursk are well-known engagements on the Eastern Front during the Second World War. But by some standards, the Battle of the Dnieper, which took place along the river of the same name in Ukraine between August and December 1943, was the largest battle in human history. Approximately four million men fought across an enormous front stretching from Smolensk in western Russia all the way to the Black Sea. During the Battle of the Dnieper, the Soviets pushed the Germans back into western Ukraine and liberated Kyiv.

While the Second World War brought death and destruction to many parts of the world, it would be hard to overstate the particular horrors it visited on Eastern Europe. Well over twenty million Soviet soldiers and citizens lost their lives between the Nazi invasion of the USSR in the summer of 1941 and the end of the war four years later. The figure probably lies closer to thirty million when associated deaths are taken into account, many of which were due to a post-war famine in 1946 and 1947 resulting from the devastation of the Soviet economy. Moreover, entire cities were reduced to rubble by the German military, and battles for cities like Leningrad, Moscow, Stalingrad, and Sevastopol led to millions of deaths. This is not to understate or belittle the sacrifice of the Western Allies, yet it must be pointed out that while Canada, for instance, lost 45,000 people in the whole war, at least ten times that many Soviets died in the Battle of Stalingrad alone, a number that may actually have been in excess of half a million. While the battle for Stalingrad is well known, few engagements capture the sheer scale of the war in the East like the strangely forgotten Battle of the Dnieper between August and December 1943.

The Battle of the Dnieper was part of the vast Soviet counter-attack along the Eastern Front in 1943. After the Germans failed to capture Moscow and Leningrad in the late autumn and early winter of 1941, the focus of Hitler's strategy in Russia shifted southwards to Stalingrad, where Hitler hoped to capture the city and, with it, control of the oilfields of the Caucasus region. Doing so would guarantee vital supplies of oil for the German military and economy and would also deprive the Soviets of the same. The Battle of Stalingrad occurred between August 1942 and February 1943. At the end of it, the Soviets had encircled the German 6th Army, and after a long siege of their position, during which tens of thousands of German soldiers starved to death, the remaining 91,000 German men were captured.

In retrospect, the outcome of the war was clear after Stalingrad. Barring a major falling out between the Allies, or Germany developing a weapon of mass destruction–which, as we have already seen, had become very unlikely–there was no doubt that the Nazis would ultimately be defeated. But it would take another two and a half years of senseless slaughter before the war would be brought to an end in Berlin. In the meantime, after their victory at Stalingrad, the Soviets began their counter-attack to push the Germans, Romanians, and other Axis forces westwards. In July and August of 1943, this effort focused on the Battle of Kursk across the Kursk region of southwestern Russia, an engagement that became the largest tank battle in military history. Kursk is often discussed in histories of the war, but far less noted is the battle which began immediately after Kursk, the Battle of the Dnieper in eastern and central Ukraine.

The Dnieper, or Dnipro, River rises in Belarus near the city of Smolensk and flows south, into Ukraine and beyond, until it pours out into the Black Sea at Kherson. There are many large towns and cities built along the course of the Dnieper, the fourth largest river in Europe: notably Kyiv, but also Dnipro and Zaporizhzhia. In places, the Dnieper is a very wide river, with some areas having been widened further by the creation of artificial reservoirs; to the north of Kyiv, for instance, the river is several kilometers wide in places. This meant that in 1943, when the Soviets began their plans to campaign westwards, they had to attempt to secure the bridgeheads over the river or else find fordable crossing points. In reality, the Germans tried to destroy the bridgeheads as they retreated, resulting in the Russians having to secure the eastern banks of the river before regrouping and attempting to progress onward once the river was fully secured. Broadly speaking, the Battle of the Dnieper focused on capturing Kyiv and other cities and towns along the river.

The Battle of the Dnieper was unquestionably one of the largest engagements of the entire war. Over the course of four months, it involved upwards of four million soldiers and military personnel, with just under three-quarters of these fighting on the Soviet side, and slightly less than a million Germans and Romanians, along with small contingents of other Axis troops, bringing the Axis forces to nearly a million men at one point, counting reinforcements. Erich von Manstein, arguably the most accomplished German general of the war, who had devised much of the plan for the invasion of France and the Low Countries in 1940 and who oversaw the victory in the Battle of Kerch Peninsula in 1942, was the overall German commander. Under von Manstein's control were the German 8th Army, a re-established 6th Army, parts of the 2nd Army, and the 1st and 4th Panzer armies. Hermann Hoth, a veteran general and experienced tank commander, was in command of several of these divisions under von Manstein. The numerous Soviet forces were led by Marshal Georgy Khukov, who was in overall operational control, but others–such as Nikolai Vatutin, Ivan Konev, and Konstantin Rokossovsky–commanded large divisions of the Soviet armies.

The Battle of the Dnieper was fought along a nearly 1,400-kilometer front stretching from Smolensk in the north all the way to an area near Rostov and the Sea of Azov in the south. In August and September 1943, the Soviets sought to secure the northern end of the Crimean Peninsula and isolate the German forces along the northern shore of the Sea of Azov, along with striking from Belgorod towards Kyiv, several hundred kilometers to the west. Following the Battle of Kursk that summer, the German supply of functional tanks was greatly reduced, and for the first time on the Eastern Front, the Soviets had a major tactical advantage, outnumbering the Germans in tanks and planes by two-and-a-half to one. Because of this, the first two to three weeks of the assault went well, and by mid-September, von Manstein was overseeing a controlled withdrawal to the Dnieper, where a stouter resistance was made possible by taking advantage of the river and the geography of central Ukraine.

Much of the fighting in October and November was concentrated around the city of Kyiv. In October, Hoth's 4th Panzer

THE BATTLE OF THE DNIEPER

Army and elements of the other German forces began fighting a phased withdrawal in the region to the east and south of Kyiv against three-quarters of a million Soviet soldiers under the overall command of Vatutin. For weeks, bitter fighting occurred at the Burkin bend, a southern bridgehead guarding the approach to the city. Eventually, Vatutin changed course and attempted a more northerly advance, splitting his troops in the process. The aim was to secure routes over the Dnieper, then advance westwards beyond Kyiv before turning back in an arc and effectively encircling the city. In doing this, the Soviets were using methods the Germans themselves had employed against the Red Army in the autumn of 1941, when they captured hundreds of thousands of Soviet soldiers through mass encirclements. The Russians had used the same approach to destroy the German 6th Army at Stalingrad, and Vatutin aimed to employ it again at Kyiv.

In the course of October, Vatutin's forces made progress in crossing the Dnieper to both the north and south of Kyiv. They did so through a range of inventive methods that included carrying scores of ferry boats overland and putting them into the river, as well as attempting to build bridges and barges on the Dnieper. In a few instances, Vatutin's troops are believed to have even attempted to finish construction of the tunnels under the bed of the Dnieper River that Stalin had ordered built in the late 1930s. Through these multifaceted methods, Vatutin's men managed to bring their forces over by the end of October, and in the first days of November, they circled back towards Kyiv. Fierce street fighting occurred over the next week and a half in late October and early November, during which Vatutin made substantial gains in and around Kyiv but failed to fully encircle Hoth and the 4th Panzer Army. Attritional fighting followed for weeks thereafter, reportedly complicated by striking levels of drunkenness within some of the Soviet ranks after the discovery of a large cache of German alcohol supplies near Kyiv.

Von Manstein ordered a retreat from Kyiv and the Dnieper in early December, and a gradual withdrawal was affected in the days that followed. The Battle of the Dnieper effectively came to an end in the middle of December with Kyiv, the third largest city of the USSR, falling back into Soviet hands. As ever, the Soviets had suffered far more casualties than the Germans. However, the ratio had declined considerably from where it had been in 1941 and early 1942. With better strategic methods, greater airpower, and a superior number of tanks, the Soviets only lost roughly four men for every German lost. Much of this is explained by the Germans having had the defensive advantage. In any case, the loss of well over a hundred thousand German men, with several hundred thousand more wounded or otherwise taken out of action, was catastrophic, and came at a time when German manpower was already dwindling precipitously and Hitler was ordering the conscription of men in their fifties, who had served in the First World War, to plug the gaps in the lines on the Eastern Front.

With the victory in the Battle of the Dnieper, the Soviet advance westwards in Ukraine accelerated in the first months of 1944, aided by the fact that Hitler and his generals were prioritizing the provision of troops to the fronts further to the north in Poland and the Baltic Sea region. Facing less stiff resistance in western Ukraine, the Russians, Ukrainians, and others were able to proceed into Romania by as early as March 1944, and they began a campaign against Germany's most important remaining ally in Europe, liberating much of the Eastern Balkans that autumn. Meanwhile, further north, the race was on to Berlin as the Western Allies began their own assault on France that summer.

SOURCES AND FURTHER READING:

Forczyk, Robert. 2016. *The Dnepr 1943: Hitler's Eastern Rampart Crumbles.* Oxford: Osprey Publishing.

Forczyk, Robert. 2015. *When Titans Clashes: How the Red Army Stopped Hitler.* Lawrence, KS: University Press of Kansas.

Harrison, Richard. 2018. *The Battle of the Dnieper: The Red Army›s Forcing of the East Wall, September-December 1943.* Warwick, UK: Helion and Company.

Nikolaieff, A.M. "The Red Army in the Second World War." *The Russian Review* 7, no. 1 (1947): 49-60.

Trigg, Jonathan. 2017. *Death on the Don: The Destruction of Germany's Allies on the Eastern Front.* Charleston, SC: History Press.

Operation Achse
(September 1943)

Accounts of the Second World War often overlook the fact that the Italians actually overthrew Mussolini in the early autumn of 1943 and attempted to side with the Allies. The only reason the war continued in Italy was that the Germans had prepared for this. They launched Operation Achse to take control of the Italian positions across northern and central Italy, as well as in the Balkans and the Greek islands, and prolonged the war there for another two years.

One of the many myths that prevail about the Second World War is that the Italians were still allied with the Germans and offering dogged resistance to the Allies in Italy right down to the end of the war in the late spring of 1945. This is not the case. Admittedly, some Italians in the north of Italy were still adherents of the Axis cause, but for the most part, the Italians abandoned the Germans in the late summer and early autumn of 1943, overthrowing Benito Mussolini and siding with the American and British forces that were advancing towards Rome from southern Italy. The perception that the Italians remained a core part of the Axis Powers was only created by the fact that the Germans rescued Mussolini from captivity and brought him to northern Italy to act as the figurehead of a new Italian Socialist Republic that was ruled from the small town of Salò on the shores of Lake Garda. We should not be fooled by the ruse. In reality, the Allies in Italy were fighting German forces, for the most part, as the Nazis had seized control of northern and central Italy following Mussolini's overthrow. Central to this effort was Operation Achse, which occurred in early September of 1943 and was a pivotal event in the war in Italy, though a strangely forgotten one.

Operation Achse came about due to months of developments in Italy and the Mediterranean that began in the spring of 1943. The Western Allies had hemmed the Italians and Germans in North Africa into Tunisia by that time, and they achieved victory in the campaign there, capturing over 200,000 Axis soldiers as operations ended on the 13th of May. By then, operational planning for opening a Southern Front in Italy was well underway, and Operation Husky was launched to invade Sicily just two months later, on the 11th of July. Operation Husky met with swift success, with Palermo falling to the Western Allies on the 22nd of July; by the end of July, the western half of Sicily was under Allied control, and the drive to capture the east and cross the Straits of Messina to Calabria on the Italian mainland was poised to begin.

In Rome, the Grand Council of Fascism, the leading governing body of Italian fascism, met on the night of the 24th of July, just two days after the fall of Palermo, to decide what they should do in the face of this immense crisis. After long deliberations that lasted well into the early hours of the 25th of July, a vote was held on the issue of Mussolini's leadership. The Grand Council voted heavily in favor of deposing Mussolini and elevating General Pietro Badoglio to Prime Minister of Italy, and Mussolini was immediately placed under arrest. It then appeared that a lull in activity followed for weeks, but in reality, three different processes were underway. First, the Americans, British, and Canadians were mopping up the last isolated pockets of resistance in Sicily. This was done by mid-August, and they began preparing for a large invasion of southern Italy in early September. Second, the Grand Council of Fascism in Rome began sending out diplomatic feelers to the Americans and British to discuss the terms under which Italy would abjure its alliance with Germany and join the Allies. And third and finally, the Nazis began preparing for the possibility of an Italian betrayal by planning what was initially termed Operation Alaric, after a Germanic warlord who sacked the city of Rome in the year 410, but then renamed Operation *Achse*, meaning Operation Axis. The new name suggested that the operation was designed to save the German-Italian component of the Axis alliance.

The plan for Operation Achse was to forcibly disarm the Italian fascist armies throughout Europe in one massive and coordinated swoop if and when word came that the fascist government in Rome had made an agreement with the Allies to switch sides. Planning for this contingency had begun as early as May 1943, when intelligence began arriving in Berlin about the declining morale across Italy and the growing sentiment in favor of exiting the war. Despite the Allied advances in Sicily in the summer of 1943, there were still over a million Italian soldiers in arms throughout Europe. Many of them were concentrated in Italy, but there were also large contingents in other regions that the Italians had claimed as part of their expanding empire since 1939, notably the Balkans, Greece, several islands of the Eastern Mediterranean, and parts of southern and eastern France. If the Italian government switched sides and managed to win over the majority of the commanders of these disparate military forces, it would spell disaster for the German cause in the Mediterranean theater. Consequently, the object of Operation Achse was to speedily disarm hundreds of thousands of Italian troops across

Europe if an armistice between the fascist government and the Western Allies was declared. The Germans could then potentially rearm the divisions that were deemed loyal and use them to defend central Italy against an Allied attack. Another key goal of Operation Achse was to secure control of Rome. All of this would be possible because, by mid-1943, German divisions were embedded everywhere around the Mediterranean alongside Italian divisions, to aid the militarily incompetent Italians in keeping control of Italy and other regions like Albania, Greece, and Yugoslavia. The plan had yet another potential benefit, as well, of capturing millions of weapons and thousands of tanks, planes, and other war machinery that were in increasingly short supply for the Germans by the autumn of 1943.

Operation Achse was soon initiated. The Americans and British began crossing the Straits of Messina and parts of the Tyrrhenian Sea towards Salerno and other ports in the Bay of Naples region on the 3rd of September 1943, beginning their invasion of mainland Italy. That same day, the fascist government in Rome finally reached an armistice agreement with Washington and London, under which the Allies would begin occupying southern and central Italy, using Rome as their base of operations to strike at northern Italy, which was being flooded with German divisions and which was effectively outside their sphere of control. The armistice agreement was made public five days later, on the 8th of September. As soon as the Germans learned what was afoot, Operation Achse began.

Operation Achse was both a demonstration of the success of German military planning and an indication of how low Italian morale was and how completely uncoordinated Italian military efforts were. Hundreds of thousands of Italian troops were successfully disarmed in Italy, the Balkans, Greece, the Greek islands, and southern France between the 8th and 11th of September. On the 10th, German forces entered Rome, causing the Grand Council of Fascism to flee from the capital and establish a new government, allied with the US and Britain, in the city of Brindisi in the south. Naples was also secured for the Germans, though an anti-German popular uprising there in the final days of September would lead to a quick German abandonment of the city. Henceforth, the Germans established the Gustav Line, which ran right across Italy and through the town of Cassino south of Rome, as the defensive perimeter of its possessions in central Italy, while the Allies occupied the southern part of the peninsula. The German-occupied regions would subsequently be formed into a rump state controlled by the Nazis that was called the Italian Socialist Republic to give it a veneer of legitimacy.

There were significant clashes associated with Operation Achse. In some areas, entire divisions and armies surrendered and were disarmed fairly quickly and without much struggle. However, some Italian commanders were determined to resist a German takeover and had reached the same conclusions as the members of the Grand Council of Fascism had weeks earlier, and they offered resistance. Some of this resistance was in Italy, though strangely enough, the more concerted opposition was found in the east, in the Balkans, and some of the Greek islands. The Italian commander of the port of Dubrovnik, for instance, refused to surrender, and fighting ensued before the Germans took the town. The most intense fighting of all was seen on the islands of Corfu and Cephalonia, where thousands of men died in clashes that went into mid-September. Thousands more members of the Italian garrisons were executed when the Germans finally took control of the islands around the 15th of September. Finally, in the Balkans, tens of thousands of Italian troops simply dispersed into the countryside when they learned what was happening, and many joined the Yugoslav and Albanian partisans who had been engaged in guerilla warfare against the Axis occupying powers since 1941.

In other respects, the results of Operation Achse were mixed. A substantial proportion of the Italian fleet stationed in northern Italian ports like Spezia and Genoa managed to sail south, and eventually joined the Allies in southern Italy, although some were sunk by German attacks as they sailed south towards Corsica and Sardinia. On the other hand, the Italian Air Force was completely unprepared for the events of early September, and approximately two-thirds of its 800 or so operational aircraft fell into the hands of the Germans. And once the dust settled, the Germans were able to recruit tens of thousands of Italians who had been disarmed, men who were still committed to the Axis cause, back into their ranks.

A follow-up to Operation Achse was the Gran Sasso Raid. Following the overthrow of Mussolini and the *volte-face* by the fascist government in Rome to support the Western Allies in the late summer and early autumn of 1943, Mussolini was taken to the Hotel Campo Imperatore on a mountaintop in the Gran Sasso Massif region of central Italy. There Mussolini was held prisoner, the obscure location designed to prevent the Germans from rescuing Hitler's ally. It didn't work. On the 12th of September 1943, while Operation Achse was winding down, several dozen German commandos paraglided to the top of the mountain near the hotel and broke in, rescuing Mussolini and flying him to northern Italy, where he was placed in charge of the new Italian Socialist Republic at the town of Salò and became the stooge of a Nazi puppet state. Even Churchill had to admit that the Gran Sasso Raid was a mission "of great daring."

As a consequence of the comparative success of Operation Achse and the Gran Sasso Raid, the Italian Socialist Republic and the German forces in northern and central Italy were able to offer concerted resistance to the Western Allies from the autumn of 1943 onwards. In late 1943 and well into 1944, fighting focused on capturing Rome, which was only liberated in early June 1944. By that time, the Allied focus had switched to northern France and the D-Day landings, while Italy was increasingly viewed as somewhere that could be used to launch bombing raids against Austria and southern Germany. As such, the campaign on land in Italy proceeded slowly, with Florence liberated in August 1944.

Thereafter, matters stalled entirely. Bologna, just over 120 kilometers from Florence, was not liberated until April 1945, after intense fighting in the mountain passes of this part of Italy led to a virtual termination of the northwards drive in the winter of 1944. Consequently, northern Italy was not liberated until the final weeks of the war, and Mussolini, the man who had established the first fascist state in Europe back in 1922, was killed by a mob near Lake Como on the 28th of April 1945, two days before Hitler's suicide in Berlin. Thus while Operation Achse was successful in stalling the death of Italian fascism, nothing could prevent its ultimate demise.

SOURCES AND FURTHER READING:

Bosworth, R.J.B. 2007. *Mussolini's Italy: Life Under the Fascist Dictatorship, 1915-1945.* London: Penguin Books.

Bresadola, Gianmarco. "The Legitimising Strategies of the Nazi Administration in Northern Italy: Propaganda in the *Adriatisches Küstenland.*» *Contemporary European History 13*, no. 4 (2004): 425-451.

Katz, Robert. 2003. *The Battle for Rome: The Germans, the Allies, the Partisans, and the Pope, September 1943-June 1944.* New York: Simon and Schuster.

Lamb, Richard. 1996. *War in Italy, 1943-45: A Brutal Story.* Boston: Da Capo Press.

Morgan, Philip. 2008. *The Fall of Mussolini: Italy, the Italians, and the Second World War.* Oxford: Oxford University Press.

The Battle of Narva
(February – August 1944)

The long and bitter battle to control the Narva Peninsula in Estonia involved 125,000 German troops and over half a million Soviet soldiers. The engagement slowed the Soviet advance into western Poland and eastern Germany in 1944, which had consequences for the Soviet occupation of post-war Germany. Had the Battle of Narva not slowed them down, the Soviets might have overrun Germany before the Western Allies arrived there, with serious implications for the Cold War that followed.

There were three main targets for the Germans when they invaded Russia in 1941. The foremost target was Moscow, a city they believed would signal victory against the Soviets if they captured it; a supposition that ignored the evidence of Napoleon Bonaparte's invasion of Russia in 1812, when he captured Moscow but the Russians simply refused to surrender. Another objective was Stalingrad, which became the primary target in 1942. The third objective was Leningrad (modern-day St Petersburg). In many ways, this became the forgotten siege of the Eastern Front. While the importance of Moscow and Stalingrad ensured that the Soviets and the Germans would pump resources into the battles there in 1941 and 1942, Leningrad was low enough on the list of priorities for both Hitler and Stalin by 1942 that the efforts there descended into something of a stalemate after the Germans failed to take the city in the autumn and winter of 1941. An 872-day siege ensued, with the Germans–aided by the Finns, who had decided to ally with the Nazis as a result of the Winter War of

1939–1940–trying to blockade the city into surrendering. The Soviets did not employ enough resources to break the blockade as they prioritized a wide range of targets further south around Kursk and along the Dnieper in 1943, and thus it was late January 1944 before the Germans finally withdrew from Leningrad and the siege ended. The Battle of Narva followed.

The Battle of Narva is named after both the River Narva, which flows near the border between Russia and Estonia, and the town of Narva, which lies at the mouth of the river. The town is just over 120 kilometers or so southwest of St Petersburg, and so it was inevitable that fighting would begin there not long after the German withdrawal from Leningrad in late January 1944. Furthermore, the Narva region is ideal for fighting a defensive withdrawal, with a thin neck of land lying between the Gulf of Finland in the north and Lake Peipus to the south. Lake Peipus is the fifth largest lake in Europe and cannot be easily circumvented; therefore, in order to advance into Estonia and the wider Baltic region, the Soviets would need to break through this neck of land, which is only about fifty kilometers wide, between the gulf and the lake. Additionally, the River Narva runs between Lake Peipus and the Gulf of Finland, so the Narva region presents an ideal geographical location to slow an advancing enemy and cause enormous damage if the defensive advantage is maximized. Finally, the Sinimäed Hills, three closely connected hills near Narva, gave the Germans a height from which to operate their artillery.

The Battle of Narva played out over a period of seven months between February and August 1944. The Germans had over 100,000 men under the command of General Johannes

Friebner, an ideological Nazi who, unlike some of the commanders of the German Wehrmacht, was a confirmed adherent of Hitler's cause. Friebner was placed in charge of the newly reorganized Army Group Narva in early February 1944. It was a small force compared to what Army Group North had been when it advanced through the region to try and take Leningrad two and a half years earlier, but it was comprised of ideologically driven soldiers, many of whom were recruited from the Waffen-SS, the hardline element of the Nazi paramilitary organization that ran the concentration camp system.

Army Group Narva faced a much larger Soviet force of over 200,000 men commanded by General Leonid Govorov, who had earned promotion through his role in the defense of Leningrad the previous years. The Soviet forces were augmented by divisions of the 8th Estonian Rifle Corps, a division of Estonians who had joined the Red Army in response to the Nazi occupation of the Baltic region in 1941. Govorov's forces also had a major tactical advantage in terms of tanks, aircraft, and artillery, dwarfing the Germans by four to one in these spheres. This was a symptom of how stretched the German war economy was in Central Europe. Even with two million Poles, Czechs, and others being used as slave laborers in German factories to produce tanks and other weapons of war, the Nazis simply could not produce anywhere near enough equipment to replace what was being lost on the Eastern Front and elsewhere. Furthermore, while the Germans were eventually resupplied with 20,000 or so men during the long battle, the Soviets simply pumped more and more men into the Narva front over a period of months, meaning the Soviets had numerical superiority, with over half a million men eventually serving on their side.

The Battle of Narva occurred in a number of different stages. The first stage involved an effort by the Soviets to blitz their way over the River Narva and take the Narva Isthmus between the Gulf of Finland and Lake Peipus as quickly as possible. It was believed that if the river could be secured, the Soviets would be able to encircle the town of Narva and cut it off from resupply or reinforcement, and in the process capture tens of thousands of German troops. This would then lead to a swift occupation of the rest of Estonia, and quickly force the Finns to negotiate peace terms, as they would be entirely cut off from the German lines. That, at least, was the idea, but things did not work out that way. Instead, the Soviets suffered costly losses in two weeks of intense fighting along the Narva River in the second half of February 1944. When their assault failed, the battle descended into an attritional struggle along the Narva front, compounded by Stalin's attention shifting south to focus on the campaign in Belarus and the east of Poland in the late spring and summer of 1944.

The next major offensive did not come until the summer. By this time, the German lines were significantly depleted through months of attritional fighting in the town of Narva and all along the river to the lake. Then came the order from Hitler that a significant proportion of the troops around Narva were to be withdrawn and sent south to plug gaps in the German lines in Belarus and Poland. With this, it became clear that a further withdrawal, to what was termed the Tannenberg Line along the Sinimäed Hills, would have to be affected. The beginning of the withdrawal led to a fresh Soviet offensive in the last week of July, in which Govorov's core force of over 130,000 men squared off against a greatly reduced German force of little more than 30,000 men. By this time, Felix Steiner, a Waffen-SS commander, was in overall command of the German contingent. This fresh assault led to a week of intense fighting in which the Germans surrendered the town of Narva and retreated to the Tannenberg Line, although they inflicted large numbers of casualties along the way.

The last stand came in August 1944, at the Battle of the Tannenberg Line (also known as the Battle of the Blue Hills) to the west of the town of Narva in the Sinimäed Hills. Slightly less than 25,000 Germans managed to slow the advance of a Soviet force of over 125,000 men for nearly three weeks. Securing the Sinimäed hilltops cost Govorov tens of thousands of men and over a hundred Soviet tanks, but eventually the Russians broke through and secured the Narva Isthmus by the 10th of August, bringing the Battle of Narva to an end after six months of grueling attritional war. But this was no Soviet victory. While the Germans withdrew in a relatively orderly fashion and headed to Estonia and Latvia to fight in the late autumn and early winter of 1944, the Soviets had lost tens of thousands of men, with many, many more wounded.

So costly was the Battle of Narva that the exact scale of the Soviet losses remains unclear to this day. In later Soviet accounts and books written about what was termed "The Great Patriotic War," the Battle of Narva was conspicuously omitted. This, then, is not really a forgotten battle of the Second World War, but one which the Soviets tried to deliberately excise from the historical record, embarrassed about how poorly the Red Army had performed. Had it not occurred, however, the Soviet advance westwards would have been swifter in 1944, and Berlin most likely would have fallen before May 1945. This, in turn, would likely have had implications for the manner in which the different

Allied powers occupied Germany in 1945, as well as where the borders of Soviet-controlled East Germany and Western-aligned West Germany lay for decades to come.

The Battle of Narva, although ending in defeat, was in many ways one of the last successful performances by the German Wehrmacht. The outlook for the Germans was quickly worsening. The Soviets were beginning a rapid advance through Poland, and in the winter of 1944, would begin preparing for the campaign into eastern Germany in the spring of 1945 and the siege of Berlin. In the Balkans, many of the Germans' former allies switched sides during or after the Battle of Narva, and the Finns negotiated the Moscow Armistice within weeks of the defeat at Narva, exited the war, and returned to the borders agreed to at the end of the Winter War in March 1940. However, in contrast to all of this, the Battle of Narva was a substantial strategic success as a defensive withdrawal. It stood in sharp contrast to the debacle that was unfolding for the Germans through Operation Dragoon in southern and central France, a forgotten battle that we will soon examine.

SOURCES AND FURTHER READING:

Glantz, David M. "The Soviet-German War 1941-1945: Myths and Realities." 20th Anniversary Distinguished Lecture at The Strom Thurmond Institute of Government and Public Affairs, Clemson University, October 11, 2001.

Glantz, David M. 2002. *The Battle for Leningrad: 1941-1944.* Lawrence, KS: University Press of Kansas.

Kattago, Siobhan. "Commemorating Liberation and Occupation: War Memorials Along the Road to Narva." *Journal of Baltic Studies 39*, no. 4 (2008): 431-449.

Mälksoo, Lauri . 2006. *Otto Tief and Attempt to Restore the Independence of Estonia in 1944: A Legal Appraisal.* Edited by Toomas Hiio, Meelis Maripuu, and Indrek Paavle. Tallin, Estonia: Estonian International Commission for the Investigation of Crimes Against Humanity.

McTaggart, Pat. "The Battle of Narva, 1944." In *Hitler's Army: The Evolution and Structure of German Forces,* edited by Command Magazine, 287-308. Boston: Da Capo Press, 2003.

The Battle of Kohima
(April – May 1944)

The Battle of Kohima was the Thermopylae of the Second World War. In the late spring and early summer of 1944, about 2,500 British and Indian soldiers, many of them non-combatants with virtually no training, held out against over 15,000 Japanese soldiers who were trying to launch a Japanese incursion from Burma into India. This prevented the Japanese from striking onwards towards Calcutta.

What is typically called the China-Burma-India Theatre of the Second World War played out over a large expanse of territory, primarily in what is now Myanmar (known as Burma under British rule) and in adjoining regions in India, Bangladesh, northern Thailand, and the Yunnan province along the southern Chinese border with Laos and Myanmar. The fighting there was a component of the Chinese-Japanese conflict further to the north, as the Allies were anxious to keep communication channels open with the Chinese in their capital Chongqing. Yet the core strategic concern for the Japanese was to push on from their conquest of Indochina and Malaysia, in semi-alliance with Thailand, to conquer Burma and then try to invade the northeastern corner of British India, an oil-rich and strategically important region at the time.

While much of Burma was secured by the Japanese in 1942, when their efforts were dominating the Far East, by 1943 and 1944 the war in the China-Burma-India Theatre had become a hotly contested tussle between the different sides; a tussle that was complicated by an appalling man-made famine in the Bengal region of India in 1943. The famine killed an estimated three million people and caused a breakdown in relations between the head of the Allied command in the region, General Joseph Stilwell, and the Chinese Nationalist leader, Chiang Kai-shek, over the enormous levels of corruption within the Chinese government that allowed much of the Allied financing and aid to the Chinese to be stolen by Nationalist officials. It was in this context that the Battle of Kohima occurred in the late spring and early summer of 1944.

Kohima lies in the south of the Nagaland region on the northeastern extremity of India today, the strip of land in the country that curls north of Bangladesh and back down south to the border with Myanmar. It is a small urban center by the standards of the rest of India, although it has been a notable urban settlement in Nagaland itself for over a century. At the outbreak of the Second World War, thousands of people lived in Kohima. By 1944, however, Kohima had been largely abandoned apart from military personnel, who were there due to Kohima's location less than a hundred kilometers from Burma and very near the front lines. Kohima was the site of the battle we are concerned with here, but it fits within the broader context of the Battle of Imphal.

Imphal is larger than Kohima and nearly a hundred kilometers south of it. Imphal is the capital of Manipur, the state to the south of Nagaland, and lies on a large plain surrounded by hills, a natural settlement site. In 1942 and 1943, the main British base near the Burma border was established at Imphal. In the spring of 1944, the Japanese plan was to try and launch a surprise assault on Imphal and the British positions in the wider region, including Kohima, and capture their supplies, because the Japanese supply lines had been severely compromised in Burma by this time. Japanese divisions would strike at Imphal from the south, while others would attack from the northeast, severing communications between Imphal and Kohima. At the same time, other divisions would encircle Kohima and aim to capture the British garrison there, as well as the town.

The plan to attack and seize Kohima and Imphal was devised by the Japanese commander, Lieutenant General Renya Mutaguchi, an experienced figure who had been involved in clashes on various fronts in China, the Malaysian Peninsula, and the attack on Singapore in 1942 before being appointed to the Burma-India Front. He led three divisions of Japanese troops that consisted of tens of thousands of soldiers, along with some smaller contingents of natives who had formed into the Indian National Army to fight alongside the Japanese against what they perceived as the real enemy, the colonial government of the British Raj. The 31st Division, consisting of over 15,000 men and led by Kotoku Sato, was the unit that did most of the fighting for the Japanese at Kohima. Sato considered Mutaguchi, his commanding officer, incompetent and often clashed with him. The British, in contrast, were far more united under the leadership of Lieutenant General Geoffrey Scoones, Lieutenant General Montagu Stopford, and Colonel Hugh Richards. Stopford's and Richards' forces numbered no more than about 2,500 men at maximum capacity and would be the core unit doing the fighting at Kohima, while the other divisions were held up at Imphal.

While the fighting at Kohima would not begin until April, the wider clashes in the northeastern extremity of India began weeks earlier in March, when the Japanese successfully executed the first steps of their plan: cutting the road between Kohima and Imphal and sundering communications between the two sites. Then in late March and April, they encircled and isolated Kohima, which lies on a hilltop of sorts, in preparation for the siege of the town. The Battle of Kohima began in the first days of April 1944 as the Japanese clashed with the British and Indians. By the middle of the month, the Japanese had largely surrounded the British and Indian troops and taken many of the key strategic points on the ridges around the town. Japan's expectation was that, with their position surrounded and the possibility of relief from Imphal having evaporated due to the Japanese capture of the road, the British would quickly surrender. It did not work out that way, however.

When the Japanese arrived at Kohima, Colonel Richards had organized a garrison of about 2,500 men, but many of these were noncombatants whose training consisted of little more than being shown how to fire their rifles as the Japanese were closing in on the town. Those who were in regiments were elements from the 161st Indian Brigade, the Queen's Own Royal West Kent Regiment, and some of the Assam Regiment. These soldiers squared off against 15,000 or more Japanese men, maintaining a tight defensive perimeter and, over a period of several weeks, enduring some of the most brutal close-quarters combat experienced anywhere in the world during the Second World War. For much of the fighting, Richards and his men had a defensive perimeter that was just 350 square meters, while the opposing forces were effectively entrenched on the two sides of what had been Kohima's tennis court. Not for nothing would this effort later be referred to as "the Thermopylae of the war," a reference to the Second Persian Invasion of Greece in the early fifth century BCE when 300 Spartan warriors held off tens of thousands of Persians at the Pass of Thermopylae in northern Greece for a significant period of time.

The Japanese had expected a capitulation after two weeks of intense daily assaults in the second half of April 1944, but none was forthcoming, and Reynolds and his men continued to hold out. It would have made for a strange scene, with Japanese soldiers scarpering across a semi-obliterated tennis court to try and break through the perimeter of a small part of Kohima in which over two thousand British and Indian soldiers and raw recruits continued to defy them. What changed matters in the end was the evolving situation elsewhere. The Japanese had counted on the fact that the British at Kohima could not be resupplied or relieved through reinforcement, yet as the days ticked by into May 1944, the strategic situation evolved further south. British reinforcements arrived from the west by train, and they eventually managed to break through the Japanese roadblocks on the road to Kohima and proceed north, where they managed to get through to reinforce Richards and his beleaguered Spartans. The Japanese decided to withdraw after several more days of fighting. By that time, the Japanese were believed to have lost some 4,000 men, while the British and Indians had suffered hundreds of casualties but held out for several weeks against a force at least six times larger than their own.

When the Japanese finally withdrew from Kohima in mid-May, the focus of the fighting switched to Imphal, and

Richards and Stopford took what was left of their units south from Kohima to aid the effort there. The Japanese still had numerical superiority in the region, but by the start of June, the superior British communications and supply lines were beginning to tell as reinforcements arrived, some airlifted in. During the battles at Kohima and Imphal and the region in between, 19,000 tons of ammunition, weapons, and other supplies were dropped into the British bases by the Royal Air Force. This allowed the existing forces there to stave off the attacks for long enough that large numbers of reinforcements could arrive in the region and tip the battle back in favor of the Allies. Conversely, the Japanese were under-supplied and their commander, Mutaguchi, foolishly maintained their attacks to the point that his men were beginning to starve before he withdrew into Burma at the beginning of July. By then, Mutaguchi had lost tens of thousands of men across the front.

The defeats along the Imphal-Kohima Front that summer not only stopped the Japanese advance, but also put the British in a position to counterattack. A new offensive began in the autumn with a push towards the city of Mandalay, with the goal of securing northern Burma and pushing the Japanese south towards Thailand and the Malaysian Peninsula. Progress was slow and tough owing to the monsoon season, but over a long campaign, the British and their allies, which included important divisions of Australians on this front, finally managed to capture Mandalay in March 1945. Rangoon was liberated in early May 1945, just as the war in Europe was coming to an end. It was not until September 1945, however, after the Japanese surrender, that Singapore was finally restored to British control. Nevertheless, the

THE BATTLE OF KOHIMA

Battle of Kohima and the wider clash in the Kohima-Imphal region in the late spring and early summer of 1944 stopped a Japanese incursion into India against striking odds and allowed the Allies to go on the offensive against the Japanese in the India-Burma Front that autumn.

SOURCES AND FURTHER READING:

Ethirajan, Anbarasan. "Kohima: Britain's 'Forgotten' Battle That Changed the Course of WWII." BBC News, February 14, 2021. https://www.bbc.com/news/world-asia-india-55625447.

Kolakowski, Christopher L. "'Is That the End or Do We Go On?': The Battle of Kohima, 1944." *Army History 111*, no. Spring (2019): 6-19.

Pal, Ranjan. "Revisiting India's Forgotten Battle of WWII: Kohima-Imphal, the Stalingrad of the East." CNN, October 4, 2020. https://www.cnn.com/travel/article/wwii-kohima-imphal-india-battle-intl-hnk/index.html.

Toland, John. 2003. *The Rising Sun: The Decline and Fall of the Japanese Empire, 1936–1945*. New York: Modern Library.

Walker, Lydia. "Why We Have Forgotten One of WWII's Most Important Battles? The Battle of Kohima Has Much to Teach Us about how We Remember the past." *The Washington Post,* June 22, 2019. https://www.washingtonpost.com/outlook/2019/06/22/why-have-we-forgotten-one-wwiis-most-important-battles/.

The Battle of Hengyang
(21 June – 8 August 1944)

Though it is often almost entirely omitted from the history of the Second World War in many countries today, it is important to remember that, other than the Eastern Front, no theater of the war was bloodier than the Chinese Front, where the Chinese Nationalists and Communists had been at war with the Japanese since 1937. The Battle of Hengyang in Hunan province has often been termed "the Stalingrad of the Chinese theater."

It has been argued with some justification that–at least when viewed from the Asian perspective–the Second World War actually started in 1937, the year the Empire of Japan invaded China and occupied a great many of that country's foremost cities in the east along the course of the Pearl, Yellow, and Yangtze Rivers and the coast of the Yellow Sea. Often termed the Second Sino-Japanese War (after the First Sino-Japanese War of 1894-1895), this conflict eventually morphed into the Second World War in December 1941, when the US and Britain went to war with Japan, and ended in the autumn of 1945, along with the wider global struggles. The argument that the Second World War should be identified as beginning in 1937, when the Japanese invaded China, is strengthened by the fact that, barring the Eastern Front in Europe, no theater of the war saw anywhere near the level of bitter fighting and mortality as the Chinese Front did. The Battle of Hengyang in the late summer of 1944 was central to the Chinese Front, and at the time, was compared to the Battle of Stalingrad in Russia.

Before exploring the Battle of Hengyang and its significance for the war, let's briefly recap the course of the war in China to 1944. The Japanese had already been in conflict with the Chinese for decades by the 1930s. In the 1890s, Japan had begun to take advantage of the fact that they had modernized much more effectively than the Chinese had. As a result, Japan was able to end Chinese dominance over the Korean Peninsula, replacing China as the main power there and eventually annexing Korea altogether in 1910. In 1931, as China was mired in the Chinese Civil War between the Nationalists (Kuomintang) and the Chinese Communists, Japan moved into the Manchuria region of northeastern China and established the puppet state of Manchukuo under Japanese control. Finally, in the summer of 1937, the government in Tokyo exploited a small border incident with the Chinese as an excuse for a more wide-ranging war and invaded China. Within months, the great cities of the east--Beijing, Shanghai, and Nanjing–all fell to the Japanese, with appalling atrocities often committed in the process.

Over the next seven years, the war between China and Japan devolved into an attritional one after these initial swift gains by the Japanese. The Chinese Nationalists and Communists retreated westwards into the interior of China and made Chongqing their new capital, establishing an uneasy alliance in the process. And while the Japanese had already conquered the preeminent Chinese cities and wanted to conquer the rest of China, it was not enough of a priority to make Japan willing to commit the necessary resources to fully defeat the Chinese in the late 1930s and early 1940s. As a result, the conflict between China and Japan began to focus on securing a chain of provinces in central China running from Hunan in the south through Hubei, Henan, Shaanxi, and Shanxi to the north. Millions of lives were lost over the years, with the death toll of the Chinese Front eventually totaling around twenty million in all. The Battle of Hengyang was one element of this much larger conflict, though a very notable one.

The Battle of Hengyang took place over a period of just under fifty days, beginning on the 21st of June and running through the 8th of August 1944. Hengyang lies in southeastern China, north of Canton and south of Hankow. Railways had been built in the 1920s and 1930s to the north and east of Hankow, and because Hengyang was located on the Canton-Hankow railway line, the city was viewed as a strategic target for anyone looking to control the many transport and communication routes through this region. Hengyang also lies at the center of a number of riverine routeways and mountain passes, as well, and thus was viewed by both the Japanese and Chinese as critical to controlling movement between southern and central China. By 1944, much of the fighting between the Japanese and the Chinese centered on this part of China as both sides sought to hold Hankow in order to either defend or attack Shanghai and Nanjing to the east; Nanjing having been the capital of the Chinese Nationalists until 1937, and site of some of the worst atrocities perpetrated by the Japanese in the whole war.

While the war in China had begun to tip in favor of the Chinese by 1944, the Japanese were the aggressors in the Battle of Hengyang. Military planners in Tokyo were concerned about a potential Chinese advance northward from Hengyang in the second half of 1944, as well as being very alarmed by the installation of air bases in southeastern China in regions China had recovered from the Japanese in the previous year. The US and British had helped China establish these air bases, leading to increased Allied air superiority over much of the South China Sea and giving American planes the ability to reach Taiwan. In response, a Japanese army of some 300,000 men was sent into Hunan province in the early summer of 1944. By mid-June, the Japanese managed to capture the city of Changsha, just over 170 kilometers to the north of Hengyang, and began marching south towards the Chinese position at Hengyang.

A contingent of the Japanese 11th Army, consisting of slightly more than 100,000 men, was at Hengyang under the command of General Isamu Yokoyama. Yokoyama was a veteran commander who was later sentenced to death–a sentence that was commuted to life in prison–for crimes committed during the war. The Chinese were in a disorganized state after the withdrawal from Changsha, which had led to chaos in the government in Chongqing as the US commander in the region, Joseph Stilwell, threatened to cut off aid to the Chinese if the Nationalists did not stop lining their own pockets with the funding provided to them from Washington. Consequently, the relatively young and inexperienced Chinese commander, Fang Xianjue, was left in charge of a paltry force of around 16,000 to 18,000 men to defend Hengyang and stop the Japanese march southward.

Over a period of seven weeks between mid-June and early August, Xianjue and his men mounted one of the most heroic defensive performances by any Chinese unit in the entire war. Outnumbered six to one, the Chinese held out against intense bombardment. They were entrenched in the city, which the civilian population had evacuated as the Japanese neared, and were soon surrounded by the Japanese. This meant that once their lines were broken through, the Chinese who remained alive would be captured–an ominous situation given that the Japanese disdain for Chinese life matched that of the Germans towards the Eastern Europeans. In addition, the Japanese had air superiority and other advantages, and were willing use to weapons such as flamethrowers and poison gas, as well as biological weapons.

The Japanese assault on the Chinese position at Hengyang came in waves. After an initial attempt to take the city in late June and the first days of July, the Japanese fell back and regrouped. A second foray occurred in mid-July, during which the Japanese attempted to take control of several elevated parts of the city perimeter that were viewed as crucial for control of the wider city, and which would allow the Japanese to install artillery and bombard the Chinese from afar. The Japanese would eventually secure several of these elevated areas, though not without costly losses as the Chinese exploited their defensive positions to the fullest. Witnesses later described Japanese bodies piled one on top of another below the heights on the 15th and 16th of July, when the Japanese finally managed to secure their targets.

Sporadic clashes followed in the second half of July, with the Chinese often engaging in urban warfare by utilizing hospitals and other large buildings as siege points. This continued into early August, but by then the situation for the Chinese, despite their valiant resistance, was dire. The Americans had airlifted in what supplies they could in the form of ammunition, weapons, food, and medical supplies, but after weeks of fighting, Fang

Xianjue and his men were short of everything, and disease and injuries were rife. As a result, Xianjue cabled back to Chongqing that he would fight to the end, if necessary, but that unless there was a relief force in the vicinity, their situation was hopeless. Unfortunately, there was no relief force, and on the 8th of August 1944, Fang offered a ceasefire on the condition that his surviving men not be mistreated. By then the Chinese were running out of bullets and had no other option. The Japanese agreed to the ceasefire. At that juncture, the Japanese had lost somewhere in the region of 20,000 to 25,000 men, with tens of thousands more wounded. Half of Fang's men, some 7,000 or so, were dead, while several thousand more were wounded. The Japanese did not honor their promise not to mistreat the POWs, though a substantial proportion of the prisoners, Fang included, later escaped from Japanese captivity and made their way back towards the Chinese lines. Fang and several of those who fought at the battle were awarded the Order of the Blue Sky and White Sun, one of the highest military honors among the Chinese Nationalists.

The Battle of Hengyang was a significant moment on the China Front in the war. Descriptions of it as the "Stalingrad of the East" are somewhat hyperbolic (millions of soldiers fought at Stalingrad), but it was quite an important battle in its own right. Although the Japanese were victorious, it was a Pyrrhic victory in which Japan lost far too many soldiers against a numerically inferior opponent. That news buoyed Chinese morale in a way that was much needed after the loss of Changsha just days before the conflict at Hengyang began. The Battle of Hengyang had altogether different implications in Tokyo. On the 22nd of July 1944, while the fighting was still underway in Hengyang, the

Prime Minister of Japan, Hideki Tojo, who is viewed as pivotal in Japan's decision to go to war with the Western Allies in 1941, resigned from his position as news arrived of the disastrous defeat at the Battles of Kohima and Imphal in India and the cost of the fighting underway at Hengyang in China. Combined with defeat in the Battle of Saipan in the Mariana Islands in mid-July, these reverses convinced many in Tokyo, and Japan at large, that the war was lost. Yet the Japanese, like the Germans in Europe, would continue to fight long beyond the point of no return.

SOURCES AND FURTHER READING:

Costello, John. 2009. *The Pacific War, 1941-1945*. New York: Harper Perennial.

Mitter, Rana. 2013. *China's War with Japan, 1937-1945: The Struggle for Survival*. London: Allen Lane.

Toland, John. 2003. *The Rising Sun: The Decline and Fall of the Japanese Empire, 1936–1945*. New York: Modern Library.

Qisheng, Wang. "17. The Battle of Hunan and the Chinese Military's Response to Operation Ichigo" In *The Battle for China: Essays on the Military History of the Sino-Japanese War of 1937-1945* edited by Mark Peattie, Edward Drea and Hans van de Ven, 403-418. Redwood City: Stanford University Press, 2010. https://doi.org/10.1515/9781503627338-028.

Xiayong, Lu, and Jiang Sheng. 2015. *Broken Dream in Hengyang City, 1944: A Complete Record of the Defense of Hengyang*. Beijing: The Great Wall Press.

The Operation Dragoon Landings
(15 August 1944)

D-Day, the Allied landings on the beaches of northern France on the 6th of June 1944 that began the liberation of Western Europe from Nazi rule, is synonymous with the Second World War in Western minds. The Operation Dragoon landings two months later in southern France, by contrast, have been almost entirely forgotten. Yet these landings paved the way for the liberation of southern, western, and central France from the Vichy collaborationist government by mid-September 1944, while also increasing the supply of war materiel into France in preparation for the invasion of western Germany.

When it comes to forgotten elements of the Second World War and battles that almost no one talks about or remembers, surely the most perplexing of all is the Allied invasion of southern France in August 1944. Initially codenamed Operation Anvil, Operation Dragoon, as it was later called, was an astonishing success. Operation Dragoon, an amphibious invasion of the French Riviera in the space of just four weeks, led to the Allied liberation of southern and central France and opened up ports like Toulon and Marseilles, allowing additional war materiel to be brought into Western Europe in preparation for the winter push into western Germany. And yet almost no one knows about Operation Dragoon today.

What is perhaps most perplexing about this is that Operation Dragoon stands in such sharp contrast to the Allied invasion of northern France just two months earlier. D-Day is lionized as the pinnacle of the Western Allies' success in the war, when a new front was finally opened against the Germans in France, leading to the liberation of Paris not long afterward. The Normandy campaign has been the subject of many films, television series, books, and even video games. Yet few people even know that Dragoon happened. Why the disparity? Admittedly, the D-Day landings in Normandy in June 1944, and the fighting in northern France and around Paris that followed, did pull most of the German resources away from places like southern France, and consequently, when Dragoon was launched in mid-August, the Allies faced a much less daunting task than had confronted the men who landed on the beaches of Normandy ten weeks earlier. It might also be said that Operation Dragoon was simply too successful, and in not offering up the kind of dramatic struggle that was seen in Normandy, simply hasn't made for good television in the same way the D-Day story has. But, as we will see, there may be another reason Dragoon has been disregarded: it was divisive, and the British were never fully happy about the strategy of the Dragoon campaign, which led many to want to downplay its significance after the war, while the virtues of D-Day were extolled.

Throughout 1943, there were debates among the Allied leaders about new fronts in the Mediterranean. Winston Churchill always had a particular view on this subject. The Greeks had been one of Britain's only allies against the Axis Powers in the dark days of late 1940 and early 1941, along with some Yugoslav partisans further to the north who held out for many months against impossible odds when Italy, and then Germany, invaded their countries. The British Prime Minister was anxious that this faithfulness to the Allied cause should be honored by helping to liberate the Balkans as soon as possible, and he was also entirely conscious that by doing so, Britain and the US might manage

THE OPERATION DRAGOON LANDINGS

to keep the Balkans within the western camp–and outside the Soviet sphere of influence–in the post-war conflict with the USSR that was already brewing. Churchill therefore pushed for opening a front in the Balkans throughout 1943, though when one was initially opened, as we have seen, it was initiated through Operation Husky in Sicily and southern Italy.

At a conference of the Allied leaders in the city of Tehran, Iran in late November and early December 1943, the issue of new fronts arose again. This was the first time that "The Big Three"--Roosevelt, Churchill, and Stalin–had met in person. Roosevelt was inclined to be far more accommodating of Stalin than Churchill was. All three agreed that 1944 would see the Western Allies open a new Western Front by invading northern France from Britain, thus spreading German resources thinner and allowing the Soviets to make swifter progress in the east. More importantly, in terms of Operation Dragoon, there was discussion of opening a separate, second front in the Mediterranean to complement the front in Italy. The Italian front was viewed as having succeeded in so far as it had knocked the Italians broadly out of the war; however, the Italian front would only have limited utility going forward, since the Alps would act as a physical barrier to any major push north of the Plain of Lombardy–if the Western Allies even got that far north. For that reason, a second front in the Mediterranean was decided upon. This was where the divide emerged, however. Churchill once again pressed the case for opening a front in the Balkans, but Roosevelt and his leading general, Eisenhower, quickly sided with Stalin and agreed that the second front should be in southern France. Thus, from the very beginning, Operation Dragoon was a contentious military endeavor.

Planning for what was initially codenamed Operation Anvil, then renamed Operation Dragoon, began early in 1944. Initially, it was intended that the invasions of both northern and southern France would take place simultaneously, a strategy which had the advantage of dividing the German forces and creating confusion as to where Axis soldiers should be sent. It was eventually decided that the operational confusion on the Allied side that might result from trying to launch two massive amphibious invasions at the same time negated the benefits of splitting the German forces in France. Therefore, it was determined that Dragoon would take place many weeks after the D-Day landings. Its primary goals were to secure the southern coast, and in particular Marseilles and Toulon. This would allow the Allies to begin landing men and war materiel safely, and proceed thereafter to liberate the bulk of what had constituted Vichy France in the south, south-west, and central parts of France.

While Operation Dragoon was delayed, it was nevertheless intended to be a large-scale affair. Over half a million soldiers arrived in Western Europe through southern France, and as with the invasion in Normandy, thousands of armored vehicles, tanks, artillery guns, and other pieces of equipment were brought in. The makeup of the forces involved in Dragoon was cosmopolitan, with the Free French comprising an even larger component than at Normandy, and divisions from the US, Britain, Canada, Australia, South Africa, and New Zealand participating. Overall command was divided between US General Jacob L. Devers and the French commander, General Jean de Lattre de Tassigny, a colorful figure who had initially accepted the Vichy collaboration with the Germans, but had subsequently gone rogue in November 1942 and ordered his

THE OPERATION DRAGOON LANDINGS

divisions to attack the Axis powers when they tried to occupy southern France following Operation Torch in North Africa. De Tassigny was then arrested and detained in a military prison in Lyon before escaping and making his way to Britain to join the Free French resistance movement. The Allied forces would face upwards of 100,000 German soldiers, primarily from the 19th Army commanded by Johannes Blaskowitz.

The beginning of the operation went almost inordinately well. On the southern French version of D-Day, the 15th of August 1944, 94,000 Allied troops landed on the beaches and other sites along the French Riviera. While four and a half thousand Allied soldiers were killed on D-Day, just 395 casualties were recorded on the day of the southern landings. The French First Army, for very understandable reasons, was anxious to take the lead in defeating Vichy France. They led the sieges of the two major southern ports, Marseille and Toulon, which were both liberated after only two weeks, on the 28th of August. The Allies had believed it would take until the end of September to secure all of the southern coast, but this deadline was met with a month to spare. There were also approximately 30,000 German soldiers taken as POWs when the southern cities were captured. In fact, the advance in the south was so successful that in the north, Paris–the ultimate goal of the Normandy campaign–was only liberated three days before Marseilles and Toulon, despite the Normandy campaign beginning over two months earlier than Dragoon.

By the time Marseilles and Toulon fell, the Germans had already begun retreating up the River Rhone into central France to avoid being encircled and cut off. German divisions were also heading eastwards into the Central Massif from cities and towns like Bordeaux, which were largely abandoned. By early September, what remained of the German 19th Army was essentially in full flight towards Alsace and Lorraine to regroup, with the rest of the German forces preparing to defend the German border against an Allied incursion later that year. Operation Dragoon ended in mid-September 1944, when divisions that had landed in southern France made contact with the US 3rd Army under General George Patton that had advanced from northern France. The whole campaign in the south had only cost the Allies around 25,000 men killed and wounded, whereas tens of thousands of German men had been captured, depriving the Nazis of many divisions that would surely have slowed the Allied advance into western Germany the following spring, had they remained free to fight. Furthermore, the Allied capture of Marseilles and Toulon allowed a growing influx of men and materiel into France.

Given all of this spectacular success, we return to the initial problem: Why has no one today heard of Operation Dragoon and the Allied invasion of southern France? The answer must surely lie in the fate of the Balkans. Because a front was not opened there by the Western Allies, the Red Army began advancing into places like Romania, Bulgaria, and Serbia and installing Soviet-aligned administrations in the autumn of 1944. For the most part, all of these countries would fall into the Soviet bloc after the war, although as we will see very shortly, the vigor displayed by the Yugoslav resistance movement in liberating Serbia and other regions continued in the post-war period, allowing Yugoslavia to hew its own path independent of Moscow. Dragoon, then, was a contradictory success. On the one hand, it was a major strategic victory, one which was more effective than the military planners had even hoped for. And yet it had long-term consequences for the Balkans and the post-war situation there. Perhaps that at least partially explains why Operation Dragoon and the second Southern Front along the French Riviera are all but forgotten today.

SOURCES AND FURTHER READING:

Devers, Jacob L. "Operation Dragoon: The Invasion of Southern France." *Military Affairs 10*, no. 2 (1946): 2-41.

Quigley, Michael T. 2016. *Operation Dragoon: The Race Up The Rhone*. Ft. Leavenworth, KS: School of Advanced Military Studies, United States Army Command and General Staff College. https://apps.dtic.mil/sti/tr/pdf/AD1022205.pdf.

Tomblin, Barbara B. 2004. *With Utmost Spirit: Allied Naval Operations in the Mediterranean, 1942-1945*. Lawrence, KS: University Press of Kentucky.

Zinsou, Cameron. "The Forgotten Story of Operation Anvil: In August 1944, the United States Executed a Gigantic Assault on Southern France. Why Does No One Remember It?" *The New York Times*, August 15, 2019.

Zinsou, Cameron. "Forgotten Fights: Operation Dragoon and the Decline of the Anglo-

American Alliance: The National WWII Museum: New Orleans." The National WWII

Museum | New Orleans, August 16, 2020. https://www.nationalww2museum.org/war/articles/operation-dragoon-anglo-american-alliance.

The Liberation of Belgrade
(15 September – 20 October 1944)

With increasing amounts of German resources and manpower committed to the fronts in Poland, France, and Italy, the Balkans became a peripheral concern in the second half of 1944, allowing Josip Broz Tito and his Yugoslav partisans to liberate the city of Belgrade. The manner in which Belgrade and the Balkans were liberated from within, rather than by the Soviets, ensured that Tito's Yugoslavia could steer its own, independent course after the war, and it did not become part of the Soviet's Warsaw Pact.

Even as the German Third Reich began to collapse in 1943 and 1944, there were remarkably few successful revolts against the Germans in the provinces. As we have seen in the discussion of Operation Achse, the Nazis were able to take over most of Italy's territories beyond those that the Western Allies already occupied in Sicily and southern Italy when the Italians decided to switch sides in the war in September 1943. The French Resistance, for all its successes in sabotaging elements of the German military machine in French ports and towns, was never able to launch a major successful revolt in France and had to wait for D-Day to begin to liberate the country after four years of occupation. And the Warsaw Uprising, the single largest revolt initiated by any occupied people against Nazi rule during the Second World War, failed after two months of vicious fighting when the Polish Home Army did not receive sufficient aid from the Soviets to the east. All of this makes the success of the Yugoslav partisans in the Balkans doubly impressive. They were, to a large extent, the only occupied people who managed to substantially liberate themselves during the war, and the ousting of the Germans from the Yugoslav capital, Belgrade, in the late autumn and early winter of 1944 was central to this effort.

The situation in the Balkans had always been difficult for the Axis Powers. Mussolini earmarked Greece and large parts of the Balkans as countries he wished to conquer as part of Italian fascism's efforts to create a latter-day Roman Empire centered on the Mediterranean, but Hitler and the Nazis had interests there, too, and wished to avoid having governments in Belgrade and Athens that were allied with the British. Hence in the spring of 1941, the Nazis undertook a giant invasion of Yugoslavia, a state which covered the modern-day countries of Serbia, Montenegro, Croatia, Bosnia and Herzegovina, Slovenia, Kosovo, and parts of northern Macedonia. A swift campaign followed, once the Germans came to the aid of the perennially lackluster Italians, and Greece was occupied by the end of April 1941, followed by an airborne invasion of Crete in May.

As swift as the conquest of the Balkans and Greece had been once the Germans intervened, the region became a hotbed of guerilla warfare and resistance between 1941 and 1944. Some groups were co-opted into the Axis cause through a strategy of favoritism the Axis used in many places; just as the Ukrainians had been favored over the Poles, the Croats were now favored over the Serbs. But the Serbs continued to resist, and launched a bitter guerilla struggle driven by Yugoslav partisans led by Josip Broz Tito, the future leader of communist Yugoslavia. Sabotage of the Axis infrastructure, attacks on smaller isolated garrisons, and eventually large attacks by thousands of partisans, turned

THE LIBERATION OF BELGRADE

Yugoslavia into one of the few places where Nazi brutality failed to totally pacify the region. A sign of the success of Tito's partisans was that by late 1943, the Germans were calling on the aid of their Bulgarian allies to invade eastern Serbia, offering land concessions there in return, while also countenancing negotiations of prisoner exchanges and potential peace with the Yugoslavs.

It was in this context that Tito and his partisans began planning to try and take Belgrade, the capital of Yugoslavia, in 1944. With every passing month, the possibility of doing so drew nearer. Axis forces were already being pulled out of the Balkans to plug gaps in the German lines in Eastern Europe in the first half of the year, and then the Western Allies invaded both northern and southern France in the summer of 1944, meaning additional Axis forces had to be sent west. Finally, in the early autumn of 1944, the Soviets began pressing into Romania and Bulgaria, defeating the Romanians and leading the Bulgarians to quickly change sides in the war. Thus, when Tito's partisans began to prepare to lay siege to Belgrade in early September 1944, they did so in the awareness that they would be aided by Soviet and Bulgarian divisions arriving from the east. By that time, the partisans were in control of large parts of the countryside and had so many volunteers wishing to join them that they had an insufficient amount of guns and ammunition to meet the demand.

The siege of Belgrade began on the 15th of September 1944. Several hundred thousand Yugoslav, Soviet, and Bulgarian troops were involved in the wider Belgrade region, though not all of these troops would be in the city, its suburbs, or even remotely close to it. Instead, many took part in the conflict in the wider Serbian region, securing the rivers, bridgeheads, and roads to the east and south of Belgrade. In the same area, a German force of over 100,000 troops was commanded by Field Marshal Maximilian von Weichs, an experienced, though hardly accomplished, general. Von Weichs had been placed in charge of Army Group F in the Balkans in 1943 in large part because he was a competent pair of hands who would oversee a staged withdrawal from Greece into Yugoslavia and then Hungary as the conditions there deteriorated for the Germans. Some of von Weichs' troops were German, others were Serb loyalists. Meanwhile, Tito was among the Yugoslav partisans closing in on Belgrade, while one of the Red Army's finest generals, Fyodor Tolbukhin, was in command of the 3rd Ukrainian Front force for the Soviets there.

After the initial foray into southern and eastern Serbia in mid-September, there was a period of intense fighting as the Yugoslavs, Soviets, and Bulgarians prepared for an assault on the city by securing control over numerous towns and arsenals along the Danube River and the main highways leading towards Belgrade. At the same time, the Germans were flying in troops from other parts of the Balkans in an effort to strengthen their position, some coming from as far afield as Greece. As a result, Athens was eventually liberated on the 12th of October, the Allies' efforts being aided by the withdrawal of German troops northwards to Belgrade. Even as all of this was occurring, Tito's partisans and their allies continued to secure major towns in eastern and southern Serbia, such as Petrovac and Negotin. By the 5th of October, the Germans were aware that Belgrade was increasingly threatened, and while continuing to prepare for the defense of the city, pulled their command center further north, out of immediate harm.

The attack on Belgrade began around the 12th of October. Mechanized units maneuvered to outflank the German troops in the city to the east of Belgrade, while other divisions advanced from the south. They managed to cut off and surround General Walter Stettner's 2nd Brandenburg Regiment, which would be engaged in bitter fighting through mid-October in an effort to break out and rejoin the German lines further north. By the 14th of October, the main defensive perimeter to the south of Belgrade had been broken through, and street fighting began. This was urban warfare in its most quintessential sense, with Yugoslav partisans and German soldiers using the urban landscape to great effect. While the fighting in Belgrade was in no way as intense as the street fighting that occurred in Stalingrad in 1942 or what would take place in Berlin in April 1945, ultimately, as more Yugoslavs, Soviets, and Bulgarians flooded into Belgrade, the Germans began withdrawing to the north and west.

Belgrade was largely liberated on the 20th of October 1944 after a five-week siege, with intense street fighting towards the end of the engagement and sporadic fighting that continued in the broader region for several weeks into November. The conclusion of the siege and the liberation of the city came after 1,287 days of German occupation. In assessing who was primarily responsible for the victory over the Nazis in Belgrade–Tito's Yugoslav partisans or the Red Army–consider that in the final stages of the street fighting in early and mid-October 1944, 2,953 members of the Yugoslav National Liberation Army lost their lives, while only 976 soldiers of the Red Army were killed.

Clearly, the Soviets and the Bulgarians aided the Yugoslavs, but the liberation of Belgrade was an act of significant self-liberation by the Yugoslav partisans, and the 20th of October has been marked as an important anniversary of the war in Yugoslavia, and now Serbia, ever since.

With the partisans' capture of Belgrade, the Germans began fighting a managed withdrawal towards the north and west of the Balkans, essentially surrendering eastern Yugoslavia to Tito and his partisans. The Serb forces expanded their numbers by offering an amnesty to Croat soldiers and others who had sided with the Nazis in recent years, and with it now inevitable that the Germans and their few remaining allies would lose the war, many accepted the offer and joined the partisans. Sarajevo became the last major Axis stronghold in the Balkans, and was not liberated until April 1945.

All of this had a substantial impact on post-war Europe. Early in 1945, Tito and his followers began establishing a new Yugoslav government, dominated by the Serbs and headed by Tito himself. This government would resume control over the Bosniaks, Croats, and other Balkan groups after the war, something which would have implications for the people of Yugoslavia through the 1990s. Also crucial was the manner in which the Yugoslav partisans liberated themselves, for the most part, and as a result were able to steer an independent line after the war. In other regions where the Soviets did the heavy lifting of liberation–Estonia, Latvia, Lithuania, Poland, Czechoslovakia, and Hungary–they imposed puppet communist governments that took their orders from Moscow and were part of the Warsaw Pact throughout the Cold War. Not so in Yugoslavia. Tito established a communist state, but it was one that hewed its own path, breaking with the Soviets in 1948 and adopting a "Third Way" that saw Yugoslavia refusing to ally with either the US or the USSR in the Cold War, and which influenced the Non-Aligned Movement in the process. It is hard to imagine the Yugoslavians having been able to do this had they not largely kicked the Axis out of Belgrade and Serbia themselves, a process in which the liberation of Belgrade in the late autumn of 1944 was the most tangible element.

SOURCES AND FURTHER READING:

Majstorovi⬛, Vojin. "The Red Army in Yugoslavia, 1944-1945." *Slavic Review* 75, no. 2 (2016): 396-421.

Tomasevich, Jozo. 2002. *War and Revolution in Yugoslavia, 1941-1945*. Stanford, CA: Stanford University Press.

Trifkovic, Gaj. ""Damned Good Amateurs": Yugoslav Partisans in the Belgrade Operation 1944." *The Journal of Slavic Military Studies* 29, no. 2 (2016): 253-278.

Ulam, Adam B. "The Background of the Soviet-Yugoslav Dispute." *The Review of Politics* 13, no. 1 (1951): 39-63.

West, Richard. 2011. *Tito and the Rise and Fall of Yugoslavia*. London: Faber and Faber.

The Battle of the Scheldt
(2 October – 8 November 1944)

The Battle of the Bulge in the Ardennes Forest is a well-known clash between the Western Allies and the Nazis on the Western Front in the months following D-Day. Just as important, however, was the forgotten Battle of the Scheldt. In securing the riverine passages between the port city of Antwerp and the North Sea for the Allies, the Battle of the Scheldt allowed both an unprecedented amount of war materiel to arrive in the Low Countries and a faster Allied advance into western Germany that winter and the following spring.

Several major battles have been fought during the twentieth century in which divisions of troops from countries that formed part of the British Empire and the subsequent British Commonwealth played a pivotal role. For instance, the Battle of Gallipoli, although an ill-conceived and ultimately costly expedition by the British against the Turks in 1915, is part of the national memory of Australia and New Zealand today, as a huge proportion of the troops in Gallipoli weren't British, but rather ANZAC (Australian and New Zealand Army Corps) soldiers who were taking part in a largely European war owing to their ties to Britain. Australian divisions such as Merrill's Marauders and the 6th Division made notable contributions to the Allied war effort in the India-Burma-China Theatre and in North Africa and the Mediterranean. But for Canada and Canadians, no engagement is more representative of their contributions to the First and Second World Wars than the Battle of the Scheldt in the late autumn of 1944.

This largely forgotten conflict focused on securing control of the lower course of the Scheldt River, which rises in northeastern France and flows through the western half of Belgium, arcing eastwards and then westwards to pour into the North Sea in the very southwestern corner of the Netherlands. Several important Belgian towns, such as Ghent, are built on its course, and even more importantly for the war effort, the port of Antwerp lies on the Scheldt River, as well. Antwerp, one of Europe's great ports, had been liberated on the 4th of September and was considered critically important for the Allies after the liberation of Paris, as tanks, armored vehicles, trucks, artillery, and all kinds of heavy machinery could be brought into Western Europe there in preparation for the planned push into western Germany in the autumn of 1944. But in order to be able to do this unhindered, the Allies needed to control the mouth of the Scheldt to ensure that traffic coming in and out of Antwerp could do so in safety. The effort to control the mouth of the river would lead to the Battle of the Scheldt.

The Battle of the Scheldt did not begin immediately after the liberation of Antwerp–and most of Belgium's other major cities and towns–in the first week of September 1944. Instead, the Allies launched Operation Market Garden, an airborne invasion of the southern Netherlands, in mid-September. The goal of Market Garden was to secure bridgeheads over the lower course of the Rhine River and allow for a larger land invasion that would liberate the Netherlands, but this operation resulted in the only major defeat suffered by the Allies in Western Europe in 1944 and 1945 when, despite liberating Eindhoven and Nijmegen for a time, they failed to secure the ultimate target, the town of Arnhem. Instead, the Allies ended

THE BATTLE OF THE SCHELDT

up pulling out of the Netherlands altogether, leaving the Netherlands under German occupation. Amsterdam and most of the Netherlands would not be liberated until the Nazis surrendered in early May of 1945. Nonetheless, the mouth of the Scheldt River had to be secured to allow Allied shipping into Antwerp. Thus, the Battle of the Scheldt was fought as a result of the failure of Operation Market Garden.

The battle began on the 2nd of October as 135,000 Allied troops pushed into northern Belgium and the southwestern corner of the Netherlands. The core of this force was the First Canadian Army led by Lieutenant-General Guy Simonds, with support from divisions of British, French, American, and Polish Resistance fighters. Overall operational control was held by a hero of some of the early British clashes with the Axis forces in North Africa in the dark days of 1940 and 1941, Field Marshal Bernard Montgomery, who had a tendency towards self-promotion, but in most practical respects, Simonds was calling the shots along the Scheldt. The Allies faced 90,000 Germans who had dug in on Walcheren Island towards the mouth of the Scheldt and the Beveland Peninsula, a strategically important region where, in 1809, the British had tried to open a bridgehead on continental Europe against the French and their Dutch allies during the Napoleonic Wars. The German forces were commanded by Gustav Adolf von Zangen, a middling general who, owing to the circumstances of the German war effort by 1944, rose to a more senior position fighting the Western Allies in France and the Low Countries from D-Day onwards. There were tens of thousands more men involved in the wider region providing support and supply to these forces, but these were the divisions that did most of the fighting.

The Battle of the Scheldt went on for five and a half weeks, into early November. In some ways, it was an unusual battle by the standards of the Second World War. The Second World War was a conflict about speed and armies moving over large battlefields and zones, the antithesis of the First World War, where millions of men fought for a few hundred kilometers of bombed-out and churned-up land in northwestern France and Belgium for four years. The Battle of the Scheldt harked back to the First World War. It was a bitter clash fought over a small span of land around the mouth of the Scheldt that became a battle of attrition. This was compounded by the late autumn and early winter weather in Northern Europe, with the North Sea blowing into the Netherlands.

Fighting was particularly intense on the 6th of October, a date known in the Canadian war memory as Black Friday. Over 140 Canadians were killed and hundreds more wounded as they tried to dislodge the Germans from the town of Woensdrecht, the first major objective of the Scheldt. A large division led by Major Douglas Chapman attacked the German lines across an open field that offered little cover, a shooting alley that was termed "The Coffin" by the Allies. Heavy German artillery and machine gun fire rained down on them from heights the Germans occupied. The fighting was effectively like the First World War, when men had to run out into No Man's Land, the empty space between the trenches on the front lines of the Western Front, looking for the next line of cover and hoping they wouldn't be hit while exposed. Chapman and his men had to fall back many times to regroup, though over time they crept slowly forward and managed to secure a better location on the periphery of Woensdrecht. It would take until the 16th of October, nearly two weeks of intense fighting, before the Germans were fully dislodged from the town.

Montgomery, who had taken a hands-off approach to the entire clash up to that point, sent more British troops to reinforce the Canadians after the fall of Woensdrecht. They spent the next two weeks clearing out German troops, who were located in small defensive locations and dug into towns and villages along the course of the Scheldt. By the end of the month, only Walcheren–which would prove difficult to take–was left in German hands. Walcheren is an island, though one which was artificially turned into a peninsula in the nineteenth century through the construction of the Sloedam that allowed a railway to run into the main town of Middelburg. As a result, it is a natural defensive point. The waters around Walcheren are too wet and marshy to cross by foot, yet too shallow for large boats carrying troops to cross, but they could be exploited in other ways, and the Allies had a plan. The British RAF bombed the waters around Walcheren, breaking the dikes and other water management systems and flooding parts of the island. This compromised some of the German defenses and also raised the water levels sufficiently for the Canadians and British to be able to float boats on the water and transport tens of thousands of troops to Walcheren. The 3rd Canadian Division, who became known as the "Water Rats"–a play on Montgomery's troops in North Africa known as the "Desert Rats"–secured Middleburg on the 6th of November. Fighting in the Battle of the Scheldt ended two days later when the Germans surrendered in large numbers, any potential retreat having been compromised by the flooding of the causeway leading off Walcheren to the north.

By the time the fighting ended, the Allies had won a major victory. They had lost nearly 13,000 men killed or wounded, over half of whom were Canadians, but over 40,000 Germans were taken prisoner at the end of the engagement, in addition to an undetermined amount of Axis casualties. It had been a difficult and costly campaign, though. The victory was very much a Canadian one, more so than any other engagement of the war, but because soldiers who fell in combat in Europe generally could not be transferred back home to North America for burial, they were laid to rest in Europe. Today, the Adegem Canadian War Cemetery in the province of Oost-Vlaanderen in Belgium, and a similar Canadian War Cemetery in Bergen-op-Zoom in the Netherlands, are the site of remembrance celebrations every year for Canada's contribution to the Second World War.

The Battle of the Scheldt was important as a critical episode in Canada's contribution to the Second World War, but it was also quite important from a strategic perspective. With the mouth of the river secured, Antwerp was open for Allied shipping to come in and out along the Scheldt once the German mines that had been planted along the riverbed were cleared, an effort that took several weeks. Once this was done, some two and a half million tons of supplies and machinery arrived in the Belgian port between November 1944 and May 1945. Hundreds of thousands of Allied soldiers who fought in Germany in 1945 did so with food in their stomachs that had come through Antwerp after the Battle of the Scheldt, and they often traveled on the backs of trucks or in tanks that had also arrived via Antwerp following the Canadian victory.

SOURCES AND FURTHER READING:

Berthiaume, Lee. "The Battle of the Scheldt: Remembering Canada's Hard-Won Victory 75 Years Later." CityNews Toronto, November 7, 2019. https://toronto.citynews.ca/2019/11/07/the-battle-of-the-scheldt-remembering-canadas-hard-won-victory-75-years-later/.

Copp, Terry. 2007. *Cinderella Amry: The Canadians in Northwest Europe, 1944-1945*. Toronto: University of Toronto Press.

Neillands, Robin. 2007. *The Battle for the Rhine 1944: Arnhem and the Ardennes: The Campaign in Europe*. London: Weidenfeld & Nicholson.

Whitaker, Dennis, and Shelagh Whitaker. 1986. *Tug of War: Eisenhower's Lost Opportunity: Allied Command & the Story Behind the Battle of the Scheldt*. New York: Beaufort Books.

Zuelhlke, Mark. 2014. *Terrible Victory: First Canadian Army and the Scheldt Estuary Campaign: September 13 - November 6, 1944*. Vancouver: Douglas & McIntyre.

The Battle of the Leyte Gulf
(23–26 October 1944)

When it comes to naval clashes of the Second World War, the Japanese attack on Pearl Harbor in December 1941, the Battle of Midway in June 1942, and the Battle of the North Atlantic between the British and the Germans are usually cited as the most significant engagements. Nevertheless, the Battle of the Leyte Gulf, between the US and Japan in the waters off the Philippines, was more important than the other battles in many ways. Though broadly forgotten today, it was by some criteria the most significant battle in naval history, and it is also notable as the last naval battle to take place between battleships in the twentieth century.

Ask anyone about the foremost clashes at sea during the Second World War and they will probably mention Pearl Harbor, which wasn't a battle so much as a demolition of parts of the US Pacific Fleet in a surprise attack, and they will likely also mention the Battle of Midway. Without a doubt, both were important. Pearl Harbor brought the United States, with all its power and might, into the Second World War, and from that point on, there really was no doubt that the Allies would be able to emerge victorious in the long run, thanks to the combined economic strength and manpower of the US and the USSR. Midway was also critical in stemming the Japanese advance in the Pacific in 1942 and making sure the US still had bridgeheads in the central Pacific it could use to launch expeditions, such as the Guadalcanal Campaign, later that year. But it is the Battle of the Leyte Gulf, an engagement that is often overlooked, that was unquestionably the largest naval battle of the Second World War. In fact, due to its scale, some military historians consider it to be the most consequential naval battle in human history.

The Battle of the Leyte Gulf came about in the late autumn of 1944, when the wider strategic situation in the Pacific Theatre was looking increasingly bleak for the Japanese. After the victories at Midway and Guadalcanal in the second half of 1942, the Americans had consolidated their hold over a wide string of smaller islands, like Wake Island and the Solomon Islands, across the Pacific Ocean in 1943. And as we have seen, in the spring and summer of 1944, the Japanese suffered a series of simultaneous defeats in the China-India-Burma Theatre. With the US war economy in full throttle by this time, new warships and transports were heading all over the Pacific and securing strongholds formerly seized by the Japanese in 1941 and 1942. Guam, for instance, was liberated at the end of July, and the Mariana Islands soon after. It was at this juncture that the long-planned liberation of the Philippines, two and a half years after the Japanese occupied the island archipelago following the Battles of Bataan and Corregidor, took place. The first major stage in the effort to liberate the Philippines was the Battle of the Leyte Gulf.

An idea of the geography of the Philippines is important to understanding this campaign. The Leyte Gulf lies east of the Philippines, an island archipelago that is basically divided into three different parts. On the northern side of the archipelago is the island of Luzon, the largest island of the Philippines, home to Manila and the site of nearly all the fighting of the Philippine Campaign of 1941 to 1942. On the southern end

THE BATTLE OF THE LEYTE GULF

of the archipelago is the second-largest Philippine island, Mindanao. In between Luzon and Mindanao is a large group of smaller islands known as the Visayas, or Middle Philippines. One of the islands on the eastern side of the Visayas is Leyte. The Leyte Gulf is on the eastern side of the island of Leyte and is a large, triangular gulf straddling eastern Leyte and the southern end of the island of Samar. Thus, it made sense for the Americans to launch their strike against the Philippines through the Leyte Gulf, a natural waterway that would allow for amphibious landings on the Visayas, as well as the eventual expansion of the campaign onto land.

The fleets involved in the Battle of the Leyte Gulf were enormous by the standards of modern naval warfare. The US 3rd and 7th Fleets were the central components of the American forces and included eight fleet carriers–the larger class of aircraft carriers–plus nine smaller light aircraft carriers and eighteen escort carriers, the ships that could launch aircraft but did not carry large numbers of them. In assessing these numbers, it is useful to note that there are only 47 aircraft carriers in the whole world in 2024. Additionally, the American armada involved in the Battle of the Leyte Gulf included twelve of the largest class of US battleships and over a hundred cruisers and destroyers, the latter of which were small, fast-moving, frigate-type ships that should not be confused with the Destroyer-class battleships of the First World War era. All of these ships were joined by hundreds of smaller motorboats and transport ships, and the whole outfit had air support from over a thousand planes, as well as additional invisible aid from US submarines. A small Australian contingent, Task Force 74, took part in the battle, too. Arrayed against this mighty force was a significantly smaller

Japanese armada, although one which still included more than half a dozen aircraft carriers, dozens of battleships, destroyers, and cruisers, and hundreds of planes. The Japanese also had the defensive position, a considerable advantage in terms of refueling and using anti-aircraft guns located on Leyte, Samar, and several other Philippine islands.

One ominous aspect of the Battle of the Leyte Gulf was that it saw the first use of *Kamikaze* aircraft. Kamikaze, meaning "divine wind," described the aircraft the Japanese had begun building out of desperation in 1944. They were little more than piloted missiles. The pilots of Kamikaze aircraft were flying suicide missions because the Japanese had developed these planes in the belief that such aircraft could cause immense damage with limited resources, ideally by crashing directly onto the deck of a battleship and sinking it. The young Japanese pilots needed to be convinced that it was worth sacrificing their lives to fly these planes, but unfortunately, far too many decided that the hollow imperial cause was worth the price. The first Kamikaze attack on a US warship in the Leyte Gulf occurred on the 25th of October 1944. Over the next nine and a half months, thousands of Kamikaze suicide missions took place, often causing immense damage, such as when the USS *Bunker Hill*, one of America's aircraft carriers, was hit near Okinawa in May 1945. The vessel didn't sink, but a major deck fire killed hundreds of US Navy personnel.

Despite the first use of the Kamikaze planes by the Japanese, the Battle of the Leyte Gulf was a daunting prospect for the heavily outnumbered Japanese. The fighting began on the 23rd of October 1944. The initial goal of the Americans was to simply cripple Japanese naval and air power around the Philippines. This would allow the Americans to secure control over the sea lanes of the South China Sea and cut Japan off from its oil supplies in the East Indies at the same time Japan was being expelled from Burma, the other source of Japanese oil. With this done, the Japanese economy and war machine would cease functioning on any major level.

The fighting kicked off with US submarine attacks on Japanese ships in the Sulu Sea, near the island of Palawan to the west of the Leyte Gulf, on the 23rd of October. At the same time, skirmishes between ships and planes were beginning to occur further east, around the Leyte Gulf. The fighting over the next three and a half days took place in four broad clashes in the waters of Leyte, Samar, and other regions near the Leyte Gulf. These were the Battle of the Sibuyan Sea, the Battle of the Surigao Strait, the Battle off Samar, and the Battle of Cape Engano. These names indicate how the fighting spilled out over a wider area than just the Leyte Gulf. Cape Engano, for instance, lies at the northern tip of the island of Luzon, nearly a thousand kilometers north of the Leyte Gulf.

The number of clashes that occurred in the four battles that make up the wider Battle of the Leyte Gulf was huge, and it would take a lengthy book to provide a detailed account of them all. In each of the clashes between the 23rd and the 26th of October, the US was on the front foot, generally speaking, and the Americans were so numerically superior to the Japanese that they were able to contain any Japanese efforts to strike at the various divisions of the American fleet. The most contentious area was the Battle off Samar, which was fought on the

25th of October. A Japanese mobile strike force managed to attack the core of the 7th Fleet there and catch them largely unaware, using air support from Luzon that included the first Kamikaze aircraft. As a result, two of the US escort carriers were sunk by the Japanese during intense fighting, in addition to several destroyers and other ships. The casualties for this single day's fighting in the Pacific were extremely high, with over 1,000 Americans losing their lives, more than three times the number lost in the Battle of Midway.

Elements of the fighting elsewhere around the Leyte Gulf were notable in other respects. For instance, on the 25th of October, the Battle of the Surigao Straits, which lies in the narrow waters between the islands of Panaon and Dinagat just slightly to the south of the Leyte Gulf, witnessed clashes between large battleships on both sides, whereas most of the fighting elsewhere involved planes attacking other planes and ships. The Japanese battleship *Fusō* was sunk during the Battle of the Surigao Straits, which was the last direct engagement between battleships during the Second World War, and also the last such clash in history, as no full-sized battleships of any major powers have engaged each other directly in armed combat since.

By the time it all ended, the Americans had won a decisive victory. Although they had suffered at least 3,000 casualties and lost two escort carriers, a dozen smaller ships, and several hundred planes, the US had crippled the Japanese navy. Twenty-eight Japanese ships were sunk, including four aircraft carriers. These were smaller aircraft carriers, but by October 1944, they constituted a large proportion of the core components of what was left of the Japanese fleet. The Japanese lost slightly more planes than the Americans did, as well, and over 10,000 Japanese men were killed or wounded during the sinking of smaller ships. This was the largest naval victory won by any power in the course of the First or Second World War, and fittingly, some of the participating American battleships were vessels that had been repaired after the attack on Pacific Harbor three years earlier.

Victory in the Battle of the Leyte Gulf did not lead to any immediate or broad land invasion across the Philippines, as there was no pressing need for a land invasion at this stage in the war. Instead, the island of Mindoro was liberated in December, and then a gradual campaign was undertaken to recapture the rest of the archipelago. Manila, for instance, was not liberated until early March of 1945. But the damage was done to the Japanese long before then; the victory in the Leyte Gulf gave the US naval and air superiority in the wider region, and the US began cutting off Japan's supply of oil from the East Indies while also bombing Japanese military installations in Indochina and southern China from the South China Sea. However, there was no talk of surrender in Tokyo. It would take a nuclear weapon before that would occur.

Does the Battle of the Leyte Gulf deserve to be called the most important naval battle in history, as some historians have argued? Quite possibly, though there are other contenders. The Battle of Actium in 31 BCE, for instance, involved hundreds of galleys and decided the final civil war of the Roman Republic in favor of Julius Caesar's grandnephew, Octavian. Following the Battle of Actium, Octavian's rival, Mark Antony, and his lover, Cleopatra, committed suicide, Octavian adopted the name Caesar Augustus, and four years later, he became the first Roman Emperor. Another contender is the Battle of Lepanto, fought between hundreds of Ottoman Empire galleys against the Spanish, Venetians, and other Italian powers in 1571, a clash in which over 400 warships took part. Oddly, both the Battle of Actium and the Battle of Lepanto took place in the same waters at the mouth of the Gulf of Corinth in western Greece. The Battle of the Leyte Gulf did not involve as many ships as Actium or Lepanto, but twentieth-century battleships and aircraft carriers are obviously more complex warships than Mediterranean galleys, so while the debate may persist, in terms of sheer strategic significance in obliterating Japanese naval power during the Second World War, the Battle of the Leyte Gulf really has no rivals for the title of most consequential naval battle in modern history. How peculiar that it is overshadowed by the Battle of Midway!

SOURCES AND FURTHER READING:

Costello, John. 2009. *The Pacific War, 1941-1945*. New York: Harper Perennial.

Friedman, Kenneth. 1999. *Afternoon of the Rising Sun: The Battle of Leyte Gulf*. Novato, CA: Presidio Press.

Hone, Trent. "U.S. Navy Surface Battle Doctrine and Victory in the Pacific." *Naval War College Review 62*, no. 1 (2009): 67-106.

Hoyt, Edwin P. 2003. *The Men of the Gambier Bay: The Amazing True Story of the Battle of Leyte Gulf*. Essex, CT: Lyons Press.

Toland, John. 2003. *The Rising Sun: The Decline and Fall of the Japanese Empire, 1936-1945*. New York: Modern Library.

The Battle of Mutanchiang
(12–16 August 1945)

> *The atomic bombs dropped on Hiroshima and Nagasaki by the US on the 6th and 9th of August 1945 are usually viewed as the final acts of the Second World War, but the Soviet invasion of Japanese Manchuria continued in the week that followed, and the Battle of Mutanchiang was actually the last official major military engagement of the war, while sporadic fighting continued in other regions into September 1945.*

An often-overlooked aspect of the Second World War is that not all of the Allied Powers were at war against the Axis Powers at the same time. While the Japanese effectively declared war on Britain and the US at the same time by attacking the British colony of Hong Kong within hours of attacking Pearl Harbor, for the most part, the Soviet Union and the Empire of Japan were not at war with one another between 1941 and 1945, despite the fact that the other major Allied and Axis powers were at war during those years.

There were good reasons for this. The Japanese had no reason to attack the Soviet Union in December 1941. Emperor Hirohito's government wanted to expand to the south, west, and east, but there was little to be gained by going north into eastern Russia, which would have pulled vital Japanese resources away from the clashes in Southeast Asia, the Pacific, and the Burma-India-China sphere. Similarly, the Soviets had enough to contend with in the winter of 1941 as the war against the Germans on the Eastern Front became the most cataclysmic war front in human history, with millions dying and entire cities in ruins. In this context, Stalin could not endure sending hundreds of thousands of troops, supported by tanks and planes, to eastern Russia, so he convinced Roosevelt and Churchill that the Soviet Union would remain out of the war in the Pacific until the Germans and Italians were defeated in Europe. Once Nazism was defeated in Europe, Stalin assured Washington, the USSR would declare war on the Japanese within three months. Stalin kept his word, and on the 8th of August 1945, exactly three months after VE (Victory in Europe) Day, the Russians declared war on the Empire of Japan.

These events are almost always overlooked in the broader history of the Second World War. There is no great mystery as to why. The first nuclear weapon was dropped on the city of Hiroshima, Japan by the US two days prior to the Soviet declaration of war, and the bombing of Nagasaki came just a day after the Russian declaration. Given that it was the deployment of the atomic bombs that finally caused the emperor and his ministers in Tokyo to conclude that they would have to surrender, the military campaign of the Soviets against Japan is typically overlooked. Yet it was significant. On the 9th of August, even as the Fat Man nuclear weapon was loaded onto the *Bockscar* B-29 bomber for its mission to Nagasaki, over a million Soviet soldiers were crossing the border into the Japanese puppet state of Manchukuo in Manchuria in northeastern China.

The Manchuria offensive would last for one week. The most consequential engagement within it was unquestionably the Battle of Mutanchiang. The city of Mudanjiang, to use the more modern rendering, is not far from the tripartite border between

THE BATTLE OF MUTANCHIANG

China, Russia, and North Korea. It lies just over 200 kilometers northwest of Vladivostok and was a major strategic target for the Soviets in August 1945. The Russian Far Eastern Front army that approached Mutanchiang and laid siege to it from the 12th of August onwards outnumbered and outmatched the Japanese in every conceivable way, as the Japanese military was in a pitiful state by the early autumn of 1945. Nearly 300,000 Soviet soldiers were involved, supported by over a thousand tanks and armored vehicles, and aided by an enormous array of artillery and siege weaponry, although Soviet airfields were limited in the east, which curtailed Russia's ability to exploit its air superiority. Not many more than 50,000 Japanese soldiers fought against Russia. They had few artillery guns or tanks, little air support, and lacked virtually everything but rifles and machine guns.

The fighting began well away from the city of Mutanchiang, where the Japanese had established a wide defensive perimeter, on the 12th of August. The Soviets had hand-picked some of their most effective fighting units from Europe and transferred them to the east to form the core of the Far Eastern Front, and these troops proved very capable. The Japanese had to begin falling back almost immediately, and by the 14th, the Japanese were retreating to the suburbs of Mutanchiang. They began to encounter stiffer resistance from the Soviets, and the Red Army commander of the Far Eastern Front, Marshal Kirill Meretskov, considered simply bypassing Mutanchiang altogether and leaving the Japanese entrenched there to move on to other targets.

This plan was abandoned on the 15th, and extra Soviet divisions were brought up to Mutanchiang to make a major push into the city. All of this was occurring even as Emperor Hirohito took to the airwaves at noon to announce to the Japanese people, most of whom were hearing the emperor's voice for the first time, that the Empire of Japan was surrendering as a result of the nuclear weapons used against Hiroshima and Nagasaki. Nevertheless, the Soviets commenced an enormous rocket and artillery attack on the Japanese positions in Mutanchiang shortly after dawn on the 16th of August and moved into the city center just before midday. Intense fighting from building to building followed as the Japanese initially refused to surrender and used the urban landscape as efficiently as possible. It was only many hours later, as news of Hirohito's announcement the previous day began filtering through the Japanese forces, that the Japanese troops began laying down their weapons and surrendering.

The exact time the fighting ended is debated. Most accounts point to late on the 16th of August 1945 as the end, while others argue that fighting persisted into the early hours of the 17th. What is clear is that in the wider clashes between the Russians and Japanese, sporadic fighting was still occurring on the islands of the Northwestern Pacific until early September. There were clear reasons for this. Even with a ceasefire agreement in place, the Soviets wanted to occupy as much territory as they could. And the Japanese also protracted the struggle. Some Japanese soldiers were determined to continue fighting without surrender. In fact, some of the small garrisons on various Pacific islands were so isolated that soldiers there continued to hold out until the late 1940s and the early 1950s. In fact, the last Japanese soldier to surrender in the Second World War, Lieutenant Hiroo Onoda, did not do so until 1974, having continued to carry out guerilla attacks from jungles in the Philippines for twenty-nine years after the war officially ended.

THE BATTLE OF MUTANCHIANG

The Battle of Mutanchiang and the entire invasion of Manchuria were central to the bloody end to the war. Approximately 12,000 Russian troops died in the wider campaign, a large number of these falling at Mutanchiang. The number of Japanese deaths was much higher; over 80,000 Japanese died, many of those deaths owing to Soviet atrocities, although the Red Army would certainly have less cause for recrimination against the Japanese than they had against the Germans in Europe. Of these 80,000 Japanese casualties, roughly 10,000 occurred at Mutanchiang. While Soviet bitterness about the invasion of the USSR and the death of tens of millions of Soviet soldiers and citizens led the Red Army to engage in mass rape and murder in Poland and Germany in 1944 and 1945, things were not as brutal at Mutanchiang. However, since the end of the Cold War and the opening of Soviet archives, it has been revealed that tens of thousands of Japanese POWs in Soviet prison camps–many of them captured at Mutanchiang and throughout Manchuria--died in captivity owing to starvation, executions, and exposure to the harsh Russian winter conditions.

One often overlooked element of this entire campaign is that, from a legal perspective, it never ended. When peace terms were established after VP (Victory in the Pacific) Day on the 15th of August, the arrangement was largely between Japan, the United States, Britain, and China. No peace settlement was ever worked out between the Soviet Union and Japan. This was more than just an oversight. There were points of contention between the Japanese and the Russians over the status of the Kuril Islands, a chain of small islands running from north of the Japanese island of Hokkaido to the southern tip of the Kamchatka Peninsula in northeastern Russia, as well as a dispute over the long, narrow

173

Sakhalin Island that lies directly north of Hokkaido and extends into the Sea of Okhotsk. These issues remained unresolved for years. It was 1956 before the Soviet-Japanese Joint Declaration formally established an armistice between the two nations, but even so, as of 2024, Russia and Japan are still theoretically at war with one another, and efforts to end this diplomatic impasse were abandoned in 2022 when Russia extended its war against Ukraine. The legacy of the Second World War is strange, indeed.

The Soviet invasion of Manchuria and the Battle of Mutanchiang might be almost entirely forgotten, but like so many of the battles and engagements charted in this book, it ought to be better remembered, for it had considerable implications for the post-war world order. The Soviet incursion there meant that the communists had a presence in northern China and northern Korea as the war came to an end, much as the Soviet liberation of Eastern Europe had consequences for countries like Poland and East Germany for decades to come as these fell into the Soviet sphere of control under the Warsaw Pact. Similarly, in the Far East, the Soviet presence in China aided the Chinese communists in the Chinese Civil War in 1949, which recommenced as soon as the Japanese were expelled from China. In Korea, a communist state under the family dictatorship of the Il-Sung family was initially sponsored by the Soviets, who had occupied the northern half of the Korean Peninsula after the Battle of Mutanchiang, and that dictatorship is still alive, although hardly well, nearly eighty years later.

As we have seen throughout this text, the legacies of the "forgotten" battles of the Second World War are still being felt today, over three-quarters of a century later. Each of the so-called forgotten battles offers valuable insights into the vagaries of war, the long-lasting and often unintended impacts of decisions made in the heat of battle, and the incalculable human cost of military conflict. We would be wise, then, to remember these battles; for while they may not be commemorated in films and novels, they helped shape the world we live in, and offer important and relevant lessons and perspectives for our collective future.

SOURCES AND FURTHER READING:

Costello, John. 2009. *The Pacific War, 1941-1945*. New York: Harper Perennial.

Drea, Edward J. "Missing Intentions: Japanese Intelligence and the Soviet Invasion of Manchuria, 1945." *Military Affairs 48*, no. 2 (1984): 66-73.

Garthoff, Raymond L. "The Soviet Manchurian Campaign, August 1945." *Military Affairs 33*, no. 2 (1969): 312-336.

Glantz, David. 2006. *Soviet Operational and Tactical Combat in Manchuria, 1945 (Soviet (Russian) Study of War)*. Oxfordshire, UK: Routledge.

Pike, Francis. "The Forgotten End of the Second World War." *The Spectator*, September 2, 2023.